D0772361

Born Arming

Born Arming

DEVELOPMENT AND MILITARY POWER IN NEW STATES

A. F. Mullins, Jr.

STANFORD UNIVERSITY PRESS 1987
Stanford, California

ISIS Studies in International Security and Arms Control

Sponsored by the Center for International Security and Arms Control
of the International Strategic Institute at Stanford

Stanford University Press

Printed in the United States of America

CIP data appear at the end of the book

For
M.F.M.

Preface

Why and how nations arm themselves are questions of great importance not only for the obvious reasons, but because of the role arms also play in the internal process of development. For the three-fourths of the world's population that live in developing countries, the second issue is often the more important. Interstate conflict has been rare in the developing world, but progress toward development has been anything but smooth and even. This research effort focuses on an investigation of the relationship between military capability and development, and addresses the questions suggested by such a relationship: What conditions promote or inhibit the acquisition of military capability? Do those conditions change with time or the geographic circumstances of individual states? And, most important, how does growth in military capability affect the rate of political and economic development? The research concentrates on a large subset of the developing world, those countries that have achieved independence in the last quarter-century and have come to statehood by a different route from most older states.

These issues are matters of more than purely theoretical interest. Although most states remain poor and militarily weak compared with the older developed states, they control large reserves of natural resources, have growing populations, and are parties to a large number of international disputes. The flow of arms transfers has kept their arsenals consistently expanding and new technology has made available to them weapons of enormous destructive power. This study attempts to examine, in a rigorous, quantitative fashion, the processes by which these new states have acquired military power, while at the same time taking into account the diversity of experience among this great variety of states. Hopefully, the results will contribute to a better understanding of the development process, of the role that military power plays in that process, and of the impact that arms transferred from the developed world have on both.

Chapter 1 reviews the theoretical and substantive issues surrounding the question of military power and development and outlines the research strategy followed. Chapter 2 reviews the military history of the new states, which in most cases extends back decades prior to independence. Chapter 3 begins the process of quantitatively examining the post-independence period, looking at the growth in military capability in the new states and their arms transactions. Chapters 4 and 5 describe the construction of an index of military capability and the results of using it to relate development to military growth. Finally, Chapter 6 summarizes the findings and examines the conclusions that can be drawn from them.

I am indebted to many people for their contributions to this book, and it is by no means possible to acknowledge all of them by name. Nevertheless, three sets of people merit special attention. First are those at the University of Michigan who helped so much in the original quantitative research; Catherine Kelleher, A. F. K. Organski, and Harold Jacobson each offered their time, their wisdom, and even their data to someone who could clearly use their counsel. Of particular help were my colleagues Richard Stoll, Glen Palmer, and especially Charles Franklin, who along with Raburn Howland and others at the Institute for Social Research helped overcome all manner of computational thickets. In the second set are my colleagues at the International Strategic Institute at Stanford, especially Chip Blacker, whose advice and careful reading of the manuscript were of great value in helping me to a sensible organization of the work. Finally, I must acknowledge my family, including the son whose misfortune it was to be born in the middle of the project, for putting up with my increasingly erratic behavior as the work moved to completion. I could not have published this book without these people, and to all I owe an immeasurable debt of gratitude. Any mistakes, of course, are entirely my own.

This book was written during a year in residence at the Stanford University Center for International Security and Arms Control, and was supported in part by a Ford Foundation grant to the Center

A.F.M., Jr.

Contents

Tables and Figures

TABLES

FIGURES

Foreword

One of the most disturbing trends in international relations since the conclusion of the Second World War has been the steady growth in the military capabilities of the less developed countries of Asia and Africa and the rising incidence of organized state-to-state violence that has accompanied this process. While the willingness of political leaders to pursue national objectives through military means is hardly unique to the Third World, it has come as something of a surprise to many students of international political and economic development that the frequency and intensity of conflict between and among less developed countries has increased despite often severe problems in sustaining economic growth.

In this study, Alden Mullins investigates the relationship between military capability and development in 46 of these countries. In a significant departure from most previous studies of this kind, Mullins argues that for many Third World countries, growth in military power and the willingness to employ it are not directly related to the state's economic performance or to the level of its economic development. In other words, some of world's poorest nations, as measured in terms of gross national product, are among the most heavily armed. Mullins concludes that it is the ready availability of weapons, manufactured in and provided by countries of the developed world, that enables these poor nations—many at the most rudimentary stages of economic development—to possess and employ modern and quite capable military arsenals.

The author also finds, in what he characterizes as a "disquieting conclusion," that the availability of weapons from such "outside" suppliers as the Soviet Union, the United States, and the countries of Western Europe will enable many Third World states to postpone indefinitely the type of economic and political development that in the past constituted the only reliable means by which nations could provide for their physical security and well-being. Given this military client-patron relationship, political leaders in many Third World countries will have few incentives to undertake the kinds of domestic reform essential to economic growth and development. In the absence of such a process, the prospects for alleviating the grinding poverty that characterizes life in much of the less developed world appear bleak. The findings of the book are enriched by a rigorous methodology that employs a sophisticated quantitative index constructed from the actual annual manpower and weapons stock totals for each country in the survey.

Born Arming: Development and Military Power in New States was written during the academic year 1985-86, while the author was in residence at the Center for International Security and Arms Control. We believe Mullins's book is an important, timely, and unique contribution to the literature on military power and political development and the Center is pleased to sponsor its publication. In this connection, we wish to express our appreciation to the Ford Foundation and Stanford University endowment that enables the Center to invite several outstanding younger scholars to the Center each year and to support their research.

Coit D. Blacker
Associate Director

Born Arming

ONE

Military Power in Developing Societies

We cannot be an independent nation
without an army of some sort.

Sylvanus Olympio
President of Togo (1960-66)

The quest for military power and its employment as a tool of national purpose have long been deemed among the principal activities of a state, and indeed the ability to employ military force is often considered a central defining property of an independent state. In many ways, the modern state is actually the product of military endeavors, whether on the part of those destined to be the kings of early modern Europe or of the democratic revolutionaries of the eighteenth- and nineteenth-century Western Hemisphere. Moreover, the relationship between the military and the state is deeper and more intimate than simply that of a servant providing through battle a territory in which the master state can organize. The internal processes of political and economic development that result in state creation have been interwoven from the very beginning with the power that military force has given rulers and the demands for resources it places on them.

For that three-fourths of the world's population living in developing countries, especially those only recently freed from colonial rule, the process of state creation is a continuing challenge, the outcome of which is by no means clear; neither is it clear what roles the military play in this process, although it has been a concern of scholars since the beginning of the decolonization period. The issue is an important one. While the military forces of most developing states, especially the new ones, are small in global terms, they operate in regions of numerous unresolved international disputes, which as the recent history of the Persian Gulf demonstrates can have global consequences. In addition, these states contain important reserves of natural resources and a significant fraction of the globe's population. The course of the development process has profound implications for the economy, ecology, and security of much more than just the developing world.

From these concerns derive the questions addressed in this book: What is the impact of the military on the development process in today's newly independent states? Is it similar to what it has been in the past? If different, what are the implications not only for the new states themselves, but for the international system as a whole? These are not original questions, for the role of the military in the development process has been assigned major importance by nearly all who have studied it. Most of the scholars who have studied military power in new states, however, have done so from the perspective of the experience of the developed world. Their research tends to focus on the division of power between military and civilian institutions and on the economic impact of defense spending on other sectors of the national economy. These are matters of great importance in developed states, which have separate and distinct military and civilian elements in their societies. But efforts to apply these concerns to the study of developing countries have produced generally less than satisfactory results, because the distinction between military and civilian authority in those countries seems misplaced, and because the indicators used in the economic analyses are inadequate to the task of properly measuring military capability and its cost.

To go beyond the perspective of these earlier studies requires recognizing that the origins of the new states of the late twentieth century are radically different from those of most of the states that preceded them. These new states are not the creations of indigenous national military forces, but rather the creations of colonial administrations. The international environment into which they were born has been very benign relative to other regions in earlier periods. The result has been a separation between the struggle for national existence and the drive for development that has significantly altered the relationship between development and military capability and the pressures on elites that go with it. To examine this altered relationship requires looking beyond the issues of military costs as a burden on developing economies or of military officers as state rulers, to see what broader inferences data about military capability can provide about the role of the armed forces in the development process. In beginning, a brief review of some earlier work on this subject will illustrate the substantial difficulties in collecting accurate data on new developing countries and serve as a prelude to the theory and design of the present research.

Civil-Military Relations in New States

In 1960, as the first of new states gained their independence, their armed forces were so weak that the likelihood of their political intervention was thoroughly discounted. Indeed, one student of the developing countries, James Coleman, went so far as to write, "Except for Sudan, none of the [sub-Sahara] African states has an army capable of exerting a political role" (in Almond & Coleman 1960: 313-14). Many authors who did see a military role were in fact optimistic about the impact of military involvement on the development process, viewing the military as an agent of change whose rational procedures, technical training, and ethic of patriotism would more than make up for the authoritarian nature of its organization. In 1962, for example, Edward Shils predicted that technically trained military officers would be impatient with traditional methods of administration and implied that they would prove to be very effective reformers (Johnson 1962: 7-68). In a study published that same year, William Gutteridge exhaustively documented the military characteristics of new defense forces and their potential as facilitators of development, with hardly any mention of the chances for or the consequences of large-scale military interventions into civilian government. Even after the first wave of military takeovers in Africa had occurred, Morris Janowitz saw the interventions as reversible and continued to expect that there would be a "civilianization" of the military as time passed (Janowitz 1964).

Hanging over all these rosy predictions, of course, was the problem of Latin America, where frequent military interventions had clearly not resulted in accelerated development. Writers like Samuel Finer (1962) and later Irving L. Horowitz (in Fidel 1975) who concentrated on this region were not so sanguine about the prospects for minimizing direct military involvement in Third World governments or of the value of the military as an agent of change. Their view that no counterweights to even small armed forces were present in most new states was borne out by the experiences of the late 1960's and the 1970's. In contrast to Europe, the professionalization of the armed forces (with its accompanying narrowed sphere of expertise and corporate loyalties) had preceded the erection of barriers between military and civilian functions (Welch 1978: 161), and military coups became increasingly frequent events among the newly independent states. Case studies of these coups or attempted coups have been quite adequate to describe the facts surrounding them and the political and ethnic histories of the regimes involved.* None, however, has proposed a comprehensive model that generalizes from the experiences to date and offers any systematic prediction of future trends. Even two very good collections (Van Doorn 1968; Janowitz & Van Doorn 1971), which fully explain the many causes and side effects of military interventions, suffer from a general tone of resignation, as though the citizens of the world's new states somehow let down all those Western scholars who had held such high hopes for them. The mostly Third World contributors to the Mowoe volume (1980) sound a more defensive note, suggesting that, for Africa at least, military coups are part of the "Africanization process," and that military rule will probably be the normal situation for much of the foreseeable future. Certainly the flash of optimism in the early 1980's (Welch 1983) over a military disengagement from politics in West Africa has not been borne out by subsequent events.

It seems clear that most of these authors, particularly in the early years, overestimated the military's impact on modernization and underestimated its willingness or ability to overthrow civilian regimes. They failed to ask the fundamental question of whether it makes sense to consider the separation of military and civil authority outside the context of developed states. In the case of developing Europe, one could make the argument that owning an army was a necessary and sometimes sufficient condition for being made king. This is not a frivolous point. Political development encompasses not only the spread of central government power for order keeping and resource extraction purposes but also the creation of a sense of identification and affiliation with that government. Prior to any significant political development, low-level administrators, regional officials, and indeed the general public owed no innate continuing loyalty to any conceptualized state or civil authority. Legitimacy accompanied political development; it did not precede it, and only as development occurred could a regime acquire authority distinct from that gained by the direct employment of military control.

With this in mind, it would have been more surprising if the less politically developed states of the twentieth century had been able to maintain what was, for them, an artificial distinction between military and civilian rule (see Welch 1985). The fact that this *is* an artificial

*See, for example, Bebler 1973 (Dahomey, Ghana, Mali, Sierra Leone); Biennen 1968 (Kenya, Sudan, Tanzania, Uganda, Zaire); Decalo 1976a (Congo-Brazzaville, Dahomey, Togo, Uganda); Gutteridge 1975 (Dahomey, Ghana, Nigeria, Sudan, Uganda, Zaire); Lefever 1970 (Ethiopia, Ghana, Zaire); and Mowoe 1980 (Algeria, Benin, Congo-Brazzaville, Ethiopia, Ghana, Mali, Nigeria, Sierra Leone, Uganda, Zaire).

distinction has tended to focus concern away from more general questions about the military and development and toward the narrow question of who is in charge.

As we have seen, the optimism about the impact of military involvement on development was derived from a belief that military personnel would be more patriotic, ethical, efficient, and technologically competent in their jobs than the average bureaucrat of civilian background. It is useful to recall Samuel Huntington's conclusion to *The Soldier and the State* (1957), in which he reflected on the order and cleanliness of the post at West Point (no billboards, no litter) and asked whether the rest of the country might not learn something of value from the military way of doing things. A decade later Kenneth Grundy (1968) noted that the African military still had a positive image in places, but that time was running out for performance to begin living up to expectations. By the 1970's, it was generally acknowledged that military personnel had no corner on methods, expertise, or standards of behavior that enabled them to outperform their civilian counterparts. Nowhere is this more evident than in Bebler's studies of Dahomey, Ghana, Mali, and Sierra Leone (1973), where it is clear that military governments rarely exercised direct administrative control below the upper levels of the bureaucracy and were, even there, susceptible to the same corruption and inefficiencies as their predecessors. In short, their performance was, on average, indistinguishable from their civilian counterparts.

J. M. Lee's analysis of the African experience is most perceptive. He argues that in Africa civil-military conflicts in the developing states reflect competition over jobs rather than policy (Lee 1969: 180). Military personnel and civilian officials outnumber the few good positions available in government and prefer, for reasons of pride or ideology, to avoid much of the foreign-owned industry being created in their countries. The military's corporate interests tend to be seen not in terms of pro-military government policies (more money for tanks or salaries, greater force levels, etc.), but in terms of prestigious roles for military personnel. In fact, Eric Nordlinger (1970), using data for 74 countries, concludes that military rulers are fundamentally unconcerned about most types of social change and generally opposed to those working for reform. Only his findings for Africa show a slight positive correlation between military rule and social and economic reform. But since many of his indicators were difficult to apply to Africa, that finding is open to question (Sarkesian in Simon 1978: 8). Robert Jackman (1976), using the same data, concludes that there is no empirical basis for assigning either progressive or reactionary labels to military governments in the Third World, and that "the simple-minded civilian-military government distinction appears to be of little use in the explanation of social change" (p. 1097).

Jackman is exactly right. We have been asking the wrong questions, from the wrong perspective. In the developed world, labeling someone a military person implies all sorts of things about his attitudes and abilities. The appropriateness of such implications is always subject to question, but this is especially so for the military of newly independent states. Military and civilian functions in new states are intermingled, and as recent research has tended to demonstrate (Bebler 1981), the previous military affiliation of a country's leaders does not appear to be the most important variable for explaining their performance. Concentration on civil-military relations has not proved to be a fruitful approach to investigating the military's role in new states. The impact, if any, of military capability on the development process does not appear to flow through the employment of military personnel in positions of administrative authority, and the search for the military-development relationship must therefore shift to other concepts.

The Economic Impact of Military Activity

A second major focus of interest for the study of military capability in developing states has been the economic impact of military activity. Research has generally concentrated on the burden that defense spending places on the budgets of these states or on the dislocations that military activity, even if funded from foreign sources, causes in the allocation of material resources and personnel. In the early 1960's, personnel dislocations seem to have been the greater concern. The military officers of the new states inherited the pay scales of their former colonial governors, who had received the same salaries as their counterparts on home service. As a result, the salaries of the new officers were very high in comparison to local wages, so high that in some countries captains made as much as cabinet ministers. Because governments were reluctant to tamper with military wages too severely and thus court trouble, military service was very attractive. Little work has been done, however, to determine whether the attraction was strong enough to result in a significant "brain drain" from the civilian sector; in fact no evidence has been presented to show any effects arising from the supposed attractiveness of military service.

The direct effect of military spending on economic activity is very difficult to investigate because such a large portion of military activity in the Third World is either directly funded or subsidized by outside powers. Benoit (1973), for example, examined the relationship of defense expenditures to the non-military economic growth rate for 44 developing countries over a fifteen-year period.* Although military spending might be a burden, his findings show a positive correlation between that burden and the growth rate. Benoit himself admits the findings are not conclusive, but he emphasizes that low growth rates have *not* been associated with heavy defense expenditures (p. 4). Frederiksen and Looney (1983), in a reanalysis of Benoit's data, attempted to demonstrate that a negative relationship between military spending and economic growth holds only for "resource constrained" countries, but Ball (1985) challenges their results, raising again the problem of accounting for foreign aid and arguing for a consistently negative relationship between spending and growth.

The difficulty in achieving a consensus in this department should not be too surprising. One would expect that significant defense expenditures would be possible only in an economy showing at least a modicum of growth and prosperity. Those who study the aggregate relationship between military spending and growth, even if they are able to subtract out the effect of foreign assistance, are still left with this difficult causal problem: if a state has to be rich to afford a lot of guns, then riches will be at least a necessary (if not sufficient) predictor of how many guns a state owns; but this does not mean that owning the guns produced the riches. The demonstration that rich states often have more guns than poor states cannot alone explain a positive relationship between military spending and growth, especially when the impact of foreign aid or foreign sales of surplus (i.e., cheap) weapons on the actual output of defense expenditures is unaccounted for.

The difficulty in accounting for outside assistance in the creation of military capability may well be a disabling problem for analyses that attempt to deal with the issue in narrow economic terms. The allocation of domestically generated resources provides only a portion, different in nearly every case, of the contributions to the military power of a developing state. Yet security, both internal and external, depends on military capability itself, the output of the expenditure process. However, since a state evaluates its security requirements on a variety of

* Benoit's indicators were defense expenditures as a percent of GDP and rate of change (in constant prices) of GDP minus defense expenditures.

geographic, demographic, and economic grounds, there is no reason to assume, *a priori*, that either wealth or growth rate is the dominant factor in its response to security problems.

Whynes (1979) attempts to measure the defense burden using as many factors as possible, including foreign assistance. Despite considerable efforts at data analysis, however, he admits "not proving . . . completely" his central thesis that defense spending, whether from domestic or foreign funding, represents a net economic cost to developing states (p. 143). Part of the problem is that it is so difficult to measure what defense spending buys, security being something that is hard to value until you do not have it, but the most intractable problem lies in interpreting the data themselves. There is little agreement on the true nature of defense burdens in the developed world, where the data are pretty good; assigning a quantitative value to defense costs in new states, where the data are woefully inaccurate, has not provided the sort of consistent results from which to make many useful inferences.

The economic burden focus, like the civil-military one, has proved inadequate to the task of fully understanding how new states are affected by the military capability acquisition process. This approach too has been trapped by a set of assumptions based on the experience of developed states. The question of costs is as narrow as the question of who is in charge. One must go beyond both to do better. To this end, we need to consider how the circumstances facing the presently developing states differ from those that faced their predecessors.

The European Experience

Early studies considered development a historically invariant phenomenon; that is, the development process was expected to take the same course in those countries currently undergoing it as it had in those that had undergone it in the past. Challenges to this view soon arose, especially from those who see the presence of already developed states as giving rise to a dependency relationship that severely restricts the prospects for development in the less-developed states. Another challenge, one more relevant to the military questions, is raised by some recent histories of Europe whose authors suggest that the military activity mandated by an anarchic international system is closely interwoven with the process of state creation and development. This view has been most broadly and persuasively put forward by William H. McNeill (1982), but others, such as Samuel Finer (in Tilly 1975), have written in the same vein. If correct, their argument means that the structure of the international system itself has a powerful impact on the way nations develop, and that (for Europe at least) the striving for military capability, far from being a drag on development, was in fact the driving force in the creation of the modern state.

At the end of the Middle Ages, improvements in military organization and tactics (especially the introduction of massed pikemen) reduced the importance of mounted knights and brought about the gradual return of massed infantry to predominance on the battlefield. As armies grew in size and sophistication after 1300, they became increasingly expensive, forcing the rulers of the European proto-states into increased penetration of their polities in order to extract the resources necessary to finance these forces. The payoff from this effort was twofold; not only did those armies created by this process handle the external security requirements, but at the same time they provided the wherewithal to achieve still greater levels of internal penetration by the central government. A feedback effect was thus obtained. Greater military power led to a greater capacity for penetration, which permitted still greater military power, and so forth.

Two assumptions (not often made explicit by historians) are required for the above model to hold for any state or system of states. First, military capability must be employable for both external and internal security. This was certainly the case during the Renaissance with the transformation of the military unit into mass disciplined infantry. Given the technological advances in armaments, the infantry unit was a far more effective battlefield weapon than the old force of mounted knights—provided it was well disciplined. Among other things, discipline implied regularly paid, and since most of the land by then had been fully distributed to feudal estates, this meant paid in cash rather than property (Howard 1976: 16-18). The ruler was thus confronted with a problem that provided its own solution. He needed money to pay his troops, but those troops, unlike the earlier knights, had no fiefs to provide a means of support on their return from battle. While the result was often problems with internal disorder, in the end more or less permanent forces were established, which (given that they were paid) owed their allegiance to the king. This meant they could be employed as agents of resource extraction in a way in which King John's barons (as they proved at Runnymede) could never have been.

The model also requires that a "law of the jungle" apply to international relations within the system in question. That is, aggrandizement by national rulers must be unfettered to the extent that only military force or the threat of force will prevent it. While some underarmed states may occasionally survive because they are buffers, or are under a patron's protection, or are geographically isolated, there can be no international norms or external powers with an interest in maintaining a balance that inhibits conflict. Rulers, therefore, have no option about defense spending. Even if they do not have territorial ambitions, they must extract increased resources just to survive within their existing borders. By this reasoning, to the extent that military force is a necessity, development is a necessity, and just as development permits the acquisition of force, so the acquisition of force permits development.

In the European model, political development is a response to external threats or opportunities. It is a pessimistic model because it implies that development is an arduous process that occurs only as a response to national security threats, and that in a peaceful world it would not occur, or at least not follow the pattern of the European experience. This implication is intuitively attractive; elites normally sit pretty well in society and are disinclined to favor change. The aristocracy of early modern Europe, however, had to tolerate the emerging middle class for the wealth it generated (to pay for the defense burden), just as its successors had to tolerate mobilizing the proletariat for the manpower it generated (to fill the mass armies).

What tie there may be between development and war is a matter of very considerable importance in our era, because many states are in the early stages of development and because there is considerable potential for conflict in the numerous unresolved issues of territory and ideology among them. If something like the European model is correct, then development will be closely related to the conflict resolution process between new states, because the military capabilities of those states are inextricably bound both to their external security requirements and to their own development histories.

A Different Experience

Although formulated to explain early modern Europe, the historical European model provides a broader and more insightful approach to researching the place of military forces in developing states than many perspectives of a more contemporary focus. The question is how the differences between modern newly independent states and their European predecessors are affecting a course of development, and how the relationship between military capability and development can be examined for the modern era while adhering to the broad scope of the original model.

To be sure, the original model was formulated to explain a period of European history that began more than 500 years ago. But comparisons between the two periods are not so farfetched as one might think, for in some respects, the present group of new states is quite similar to the earlier European ones. With the exception of Bangladesh, Nigeria, and Zaire, the populations of this group all fall between 0.5 million and 15 million. This is about the same range as for European states until 1800, when the population climbed to over 10 million in Britain, 22 million in the German states, and 27 million in France.

The armies of the two groups of states were also of similar size, at least until the Thirty Years War. There were 6,000 to 8,000 men on a side at the battle of Hastings, and by the middle of the sixteenth century, wartime armies of 40,000 were engaged in Europe, but those forces were not maintained in peacetime, when 15,000 was a large force (Tilly 1975: 101). Most of the armies of the 46 states in my study (see Table 1.1, p. 15, for a list) are peacetime forces averaging about 8,000 men. The few that have engaged in conflict tend to be much larger; Malaysia's army is 75,000 strong, South Yemen's 21,000, Somalia's 30,000, and Nigeria's 270,000.

But despite these similarities, several elements of the European model do not fit so well with the twentieth-century case. For one thing, the assumption that the same military forces acquired for external security are equally useful for internal security is probably no longer as valid as it once was. Although infantry troops likely still fit this assumption (and the armies of new states are largely light infantry), to a great extent much of modern weaponry is ill-suited to policing duties. It is hard to use a supersonic fighter to collect taxes or, as the Shah of Iran found out, to control crowds. Many states recognize this fact and maintain separate army and national police forces trained and equipped for different missions. Still, police and army units can be mutually supporting, and a large percentage of external security forces may actually be very useful as threats to compel submission to the central government. Modern internal and external security requirements still overlap at least to the extent that, by themselves, their differences should not entirely invalidate the model.

A second and more important difference between the two eras is that a "law of the jungle" clearly does not apply, at least for the first decades, to the international relations of the great majority of contemporary new states. Virtually all were created before they had any military capability for self-defense, but in every case (except Zanzibar) they survived as states through their own restraint and that of the major powers. Whether this relative state of peace can be maintained permanently is problematic. We might expect the situation to change as more states acquire the power to practice other than military restraint in their international relations. The data assembled by R. L. Butterworth and M. E. Scranton (1976), in fact, show a greater employment of military forces by new states against each other in the early 1970's (5 cases in 1972) than in the same time span a decade earlier (2 in 1962). But even if the relative security of the new states proves only temporary, it has clearly offered a breathing space for most of them lasting right up to the present.

A third and very important difference is that powers outside the system can now have a tremendous impact on the military balance within the system. The military power of the developed states is so great relative to that of the newly independent states that outside interventions can overwhelm local effort. Even as indigenous military power grows and the developed states' relative military advantage declines somewhat (making intervention more costly), the client-patron relationship will remain important through the supply of advanced military technology and the effect of military aid on the local balance of power.

These differences are a prelude to the findings of this research: the development of modern new states is proceeding differently from the course the European experience would predict because differences in the international system have altered the relationship between military power and the development process. The relatively benign international environment has reduced the external threats that drove earlier European elites to undertake development, and the existence of developed-state suppliers of military aid has often made the client-patron relationship a more important determinant of security than indigenous military preparations enabled by progress in development. The challenge for this work is to demonstrate these findings in a way that is not only quantitatively rigorous but historically comprehensive, so that credible conclusions regarding the problems and prospects of new states can be made without overgeneralizing about this enormously varied group.

The Research Design

Reviewing, the European model postulates a feedback relationship between development and military capability (Fig. 1.1). In any such two-element feedback model, four alternatives are possible, depending on whether both, neither, or either single one of the feedback relationships turns out to be specified incorrectly. In this case, if the complete feedback relationship does not hold, it may be that while development

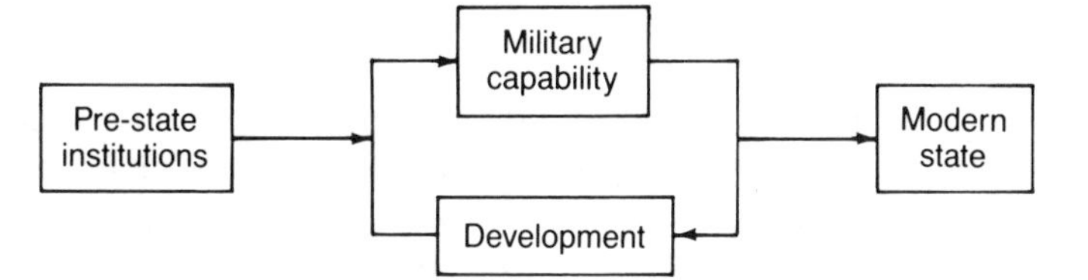

Fig. 1.1. Feedback model of military capability and development

permits the creation of military capability, it does not *require* it, that military capability *allows* development but does not *ensure* it, or that there is no relation or a negative relationship between capability and development (i.e., that neither portion of the feedback loop holds).

The matrix in Figure 1.2 displays the expected results for each of the four possible combinations of relationships between military capability and development. The combinations in squares 1 and 3 have quite negative implications for restraining arms competition, since these relationships imply that states must acquire military power to facilitate development. Number 3 is especially negative, since all states seeking to develop would need to arm, but states could acquire military capability without being successful in their development efforts. Squares 2 and 4 have more positive implications because military capability is

Relationship	Military capability required for development	Military capability not required for development
Development necessary to acquire military capability	1 Development and military growth rates necessary and sufficient predictors of each other	2 Development growth rate a necessary but not sufficient predictor of military growth rate
Development not necessary to acquire military capability	3 Military growth rate necessary but not sufficient predictor of development	4 No positive relationship between growth in military capability and development

Fig. 1.2. Matrix for development-military capability relationship

not, in these models, required for development and because, in the case of number 2, improvement in military capability is actually limited by the rate at which a state develops.

The European model's predictions are quite straightforward. Military capability requires development and at the same time itself promotes development (square 1). This relationship should be generally consistent across regions and, controlling for the level of development, across time. Testing the model provides the framework for the substantive questions of this research. What conditions promote or inhibit the acquisition of military capability? Do those conditions change with time or the geographic circumstances of individual states? Does growth in military capability itself affect the rate of political and economic development? How are the prospects for development affected by the structure of the international system itself?

The steps required to test the model are easier to describe than to implement. First, the sample must be specified; that is, the countries and time period of the study determined and the rules for making this determination described. Then a method to measure military capability must be chosen, the data assembled, and an index constructed so that an annual capability score for each state is available for cross-national and cross-temporal comparisons. Finally, the data must be analyzed to see what relationships emerge between capability and various indicators of political and economic development, in this case, paying attention to both static indicators and rates of change, and attempting to account for as many intervening and confounding factors as possible.

Two rules specify the criteria by which a state was included in this research: (1) its date of independence must have fallen between January 1, 1957, and December 31, 1981; and (2) its population at independence must have exceeded 500,000.

The 25-year time span eliminates any confounding effects of the immediate post-World War II demobilization period and holds the data-collection and manipulation work to a manageable level. Although it necessarily excludes a few states that gained independence earlier (e.g. India, Pakistan, the Philippines, Indonesia), it includes at least 85 percent of the world's new states as defined by any reasonably longer period. The independence of Ghana in 1957 really marks the beginning of the massive wave of decolonization that took place over the next decade and so marks a logical and practical starting date for the study. The 1981 cutoff was chosen not only because a quarter-century is a nice convenient figure for a survey, but also because (as discussed in the Bibliographical Note) the accuracy of data on military capability decreases as one approaches the present, and so as of this writing (1985) one can have much more confidence in data for 1981 than for more recent years.

Those so-called mini-states with fewer than 500,000 people (e.g. Barbados, Nauru, Equatorial Guinea) are excluded because they are not normally able to maintain a military force of any consequence, and their inclusion would only serve to make the search for patterns in the acquisition of military capability more difficult. Decision rules like this sometimes produce less than perfectly pleasing results (on this criterion Lesotho is included, and Swaziland is not), but it is better in a quantitative study to pick a defensible rule and stick to it than to make a series of ad hoc changes that leave the validity of the results open to challenge later. In any case, only about fifteen countries are eliminated through the use of the population criterion, mostly small island states and some African territories, leaving 46 states in Africa, South and Southeast Asia, the Middle East, and the Caribbean for the data field. Table 1.1 lists these states and their respective dates of independence.

TABLE 1.1
The 46 Countries in the Study

State	Date of independence	State	Date of independence
Algeria	1962	Malawi	1964
Angola	1975	Malaysia	1957
Bangladesh	1971	Mali	1960
Benin	1960	Mauritania	1960
Botswana	1966	Mauritius	1960
Burundi	1962	Mozambique	1975
Cameroon	1960[a]	Niger	1960
Central African Republic	1960	Nigeria	1960
Chad	1960	Papua New Guinea	1975
Congo	1960	Rwanda	1962
Cyprus	1960	Senegal	1960
Fiji	1970	Sierra Leone	1961
Gabon	1960	Singapore	1965
Ghana	1957	Somalia	1960
Guinea	1958	South Yemen	1967
Guinea-Bissau	1974	Surinam	1975
Guyana	1966	Tanzania	1961
Ivory Coast	1960	Togo	1960
Jamaica	1962	Trinidad and Tobago	1962
Kenya	1963	Uganda	1962
Kuwait	1961	Upper Volta	1960
Lesotho	1966	Zaire	1960
Malagasy	1960	Zambia	1964

[a]Formed of British Cameroon and French Cameroun; the British section did not achieve independence until 1961.

The chapters that follow outline the data assembly and analysis, examining the patterns that emerge globally and at the individual country level of analysis. As a precursor to this analysis, however, the next chapter reviews the military histories of the colonial and early post-independence period in these new states. This review may seem at first glance tangential to the major thrust of the research, but the differences between the military's role in the history of these new states and that of their European predecessors becomes clear only with this examination. It is useful in any case to note the enormous variation in circumstances and experiences across this wide sample of states as an aid in resisting the temptation, always there in the presence of a lot of data, to make inappropriate generalizations.

And, not least, history is important in reminding us that, in the end, the subjects of this research are not aggregate statistics but people, often people living and enduring in very difficult circumstances. This chapter opened with a quotation from Sylvanus Olympio to the effect that nationhood requires an army of some sort. It is appropriate to conclude the chapter by noting that in 1966 Togo suffered the first successful military coup in post-independence Africa, a coup in which President Olympio was murdered by his own troops.

TWO

The Colonial Heritage

In order to understand the behavior of the late-twentieth-century new states' military organizations and their impact on the development process, it is essential to understand their origins and the elements that made those origins unique in the world community. To that end, we begin our investigation with a short look at the pre-independence history of those military forces and the processes by which they made the transition to independence.

Imperial Forces

The military forces of nearly all modern new states were originally formed by the colonial powers as some branch of the imperial police or defense establishment. To that extent, they share a common history. But the various metropolitan powers practiced quite different policies with regard to the structure of their colonial armies, and those differences have persisted through the postcolonial period. British policy was undoubtedly influenced in greatest measure by the country's Indian experience, especially by the Sepoy mutiny of 1857 and its aftermath. A policy of racial equality and integration, which had begun to find favor in the Indian civil service just before the rebellion, was afterward forever considered out of the question by the army. Control was expected to be maintained by keeping units strictly segregated by ethnic and religious affiliations so that, in the words of one senior official of the time, "Sikh might fire into Hindu [and] Goorka into either, without any scruple in the case of need" (Gutteridge 1964: 7).

In the African context, this policy resulted in the organization of military units along ethnic lines. But troops were not randomly recruited from all ethnic groups. The British favored certain tribes over others, much as they had favored the Moslem peoples of northern India in recruiting there. In practice, this meant that the more rural or nomadic tribes came to dominate the military units of British Africa. The stated reason was that these were "warrior" tribes, better suited to combat and military life, but the effect was to keep the more educated and settled peoples (who chafed most under colonial rule) out of military service and under the guns of their ethnic rivals. When Ghana achieved independence, for example, 62 percent of its army troops were from the northern Moslem tribes, though the majority of the population was from the coastal region. Similarly, there were almost no representatives of the Kikuyu majority in Kenya's army or police (Bell in Van Doorn 1968: 260). The British found such an arrangement useful in subduing unrest, but it later had a serious negative impact on the effort to transform these colonial forces, often used as agents of suppression, into the national armies of independent states.

The British African forces were small, composed of volunteers, and generally not used for expeditions outside their own regions. They were organized into two principal commands, the Royal West African Frontier Force (RWAFF) and the King's African Rifles (KAR). Neither was used outside of Africa in the First World War, and at the start of the Second, they had only 19,000 men between them. Both were exclusively made up of infantry and logistic support units (i.e. truck drivers). Indeed, they lacked even something so basic as their own regular service units. As late as 1940, for example, the men of the RWFFA built and maintained their own barracks, and their wives cooked for them (Headrick 1978: 521).

The Second World War forced changes, and by 1945, some 525,000 Africans had been recruited (not drafted) into British service and had seen action on various fronts. But though the experience undoubtedly broadened their horizons and though the British devoted considerable effort to literacy and language instruction, these servicemen were paid less than their white counterparts, and advancement to officer rank was virtually unknown. Even after the postwar demobilization had reduced the forces by over 80 percent, the British did not draw on the great pool of African noncommissioned officers to provide commissioned officers for the colonial regiments. Although one African had been commissioned in the Gold Coast Regiment during the war, the British did not give serious consideration to using African officers in West Africa until the 1950's (Great Britain 1954). As a result, virtually all British West African armies still had a majority of British officers on independence day. East Africa trailed even further behind, with the first two African commissions via Sandhurst coming only in 1961 (Gutteridge 1964: 109). There were generally 50 or 60 Europeans in each British African battalion compared with seven or eight in India. The result was lower quality in white officers (talent spread too thinly), stifled initiative in the African soldiers, and serious difficulties in moving to a noncolonial status (Bell in Van Doorn 1968: 261).

The British seemed unable to envision Africans in control of their own existence. Nowhere is this more clear than in Hubert Moyse-Bartlett's (1956) conclusion to his exhaustive semiofficial regimental history of the King's African Rifles. After many chapters recounting the KAR's unfailing loyalty, perseverance, and courage under terrible conditions in Ethiopia and Burma, he sums up with this astonishing observation: "That they cannot be expected to attain a European standard needs no saying, but in bush and jungle warfares . . . their hardy physique, simple needs, and cheerful outlook are of outstanding value" (p. 686). A more insufferable statement of gentle condescension can scarcely be imagined.

In Malaya British recruitment policy again favored one group (Malays) over others (Chinese and Indians), but here the pattern was more like India's than Africa's (less direct British presence in the lower command structure), and although the Royal Malayan Army did not exist until 1933, by 1941 it already had 19 Malays in its officer corps. After the fall of Singapore, however, the Japanese put many of these officers to death, to the point where in 1946 the army had to be rebuilt from scratch. With the establishment of the Malay Federation, recruitment was broadened, and when the Federation Regiment was formed, it was supposed to be only 25 percent Malay (Guyot in Kelleher 1974: 34). Enlistments lagged, however, and in 1960 80 percent of the officer corps (now tested by a successful counterinsurgency war) was still Malay (Gutteridge 1964: 114).

In Britain's small island colonies, the preindependence military forces were minuscule and had exclusively internal security missions. All continued to have at least some British officers well after independence. This is somewhat surprising since the 1st West India Regiment, recruited from the Caribbean islands, had a long and exciting history of expeditionary service from the eighteenth century through the First World War, but by the 1950's only two lightly armed battalions remained. The only military force in the group to have any significance either in a domestic or an international context has been the Fijian Army, which is important in the delicate ethnic balance of power on the island and which has had extensive participation in U.N. forces. Its first units were not even established until 1949, however, and the size and experience of the current force is largely a result of the government's decision to use its army as a foreign exchange earner with the United Nations.

French colonial policy in Africa differed from British African policy in three principal respects. First, the French relied much more heavily on colonial manpower not only for local defense, but also for use overseas. The Troupes Coloniales, formally organized in 1900, were envisioned as a reserve army in the service of the French empire. The French began drafting Africans in 1912 and continued to do so right through the Second World War. At the outbreak of the war they were able to send 80,000 troops from West Africa to France, and they quickly mobilized another 90,000—and this from a population base less than one-fourth as large as the one in British West Africa. French Equatorial Africa supplied another 15,000 men (Headrick 1978: 502). In both cases, then, a much higher proportion of the population was subjected to military service than in any British possession.

The French also differed from the British in combining soldiers from every ethnic and geographic group in their units, paying little attention or respect to local customs or languages along the way. And unlike the British, they made little or no effort to give the regular soldiers a systematic education. As a result, although the French forces were far more racially integrated than the British forces on paper (colonial peoples could rise to the rank of captain in the regular French army), in practice the fraction of African officers was less than 5 percent in both armies. Gutteridge (1964: 115) estimated that in 1948 only six of 321 officers in the Equatorial region were African, and little changed before the middle 1950's.

Probably the most important lasting difference between the two colonial powers was in their treatment of African veterans. During the late 1930's, the French offered voting rights to Africans who had honorably completed their military service in an effort to ease hostility to the conscription laws. Although French African soldiers were paid less, educated less, and in many cases treated more shabbily than their British counterparts,* they had not been promised as much in the first place. Since voting privileges were virtually unobtainable for Africans except through military service, veterans wound up with a prepon-

* In one incident in 1945, Senegalese troops who had been German prisoners of war for four years staged a sitdown to protest their failure to get their back pay. The French opened fire and killed 40.

derant voice in local politics and were well able to see to their own interests. By contrast, British veterans often felt cheated in the austerity of the postwar years and had little chance to participate in home-rule efforts. In Kenya, for example, the colonial administration gave land to white British veterans who immigrated, and denied it to African veterans on the grounds that there was none available. In the words of one scholar, "What mattered in the end was not which colonial power could give the African soldier a better war but which one gave him a better peace" (Headrick 1978: 520). France's policy both before and after the war paid considerable dividends. French veterans were simply more pro-French than British veterans were pro-British. As a consequence, post-colonial France has maintained much closer military ties with the armies of the French Community than Britain has with its former colonies.

Belgian practices differed in several important respects from those of both Britain and France. The Congo region was originally subdued by Africans under European leadership; no metropolitan troops were sent to the region, and when it passed from the personal domain of the king to the government of Belgium in 1908, the organizational descendant of those first troops, the Force Publique, retained both defense and police responsibilities for the colony.

Unlike the British and French forces, the Force Publique was used very little in either world war and then only in Africa. Although it was made up of conscripts, the supply of men was left up to local tribal leaders. But by independence the benefits were such that no draftees were required to maintain the 24,000-man force. At that point, all of the private soldiers in the Force Publique were African, and all 1,100 officers were Belgian. The first African officers did not complete their Belgian educations and receive their commissions until three years after independence (Keegan 1979: 828).

The Belgians integrated their colonial forces (not only in the Congo, but also in their Trust Territories, now Rwanda and Burundi) to an even greater degree than the French. They also attempted to ensure that at least four tribes were represented in each 35-man platoon. As a result of this ethnic mixing, plus good pay and other benefits, the soldiers of the Force Publique were quite fully removed from their old tribal ties. In fact, special retirement villages were set up for veterans because they were so ill suited to life in their home villages after a military career that neither they nor their former neighbors desired their return. The effect of all this was to create a force very useful for the sometimes brutal internal security role to which it was assigned, but one with no community identification, nationalist vision, or trained leadership for the time when the Belgians would leave.

In general, the African colonial armies of Britain, France, and Belgium had been created for colonial defense and modified for internal security duty, but in the end they were well suited for neither. For decades before independence the armies of the British and French possessions were used as sources of manpower for imperial defense needs. They were predominantly light infantry forces garrisoned near embarkation ports with few weapons like tanks or aircraft suitable for defending their own borders (the colonial powers no longer fought over them) and no experience serving an indigenous regime. The Belgian troops were appropriately trained and deployed for internal security operations, but they had no African officers and no sense of loyalty to any institution other than the Force Publique itself.

The end of the colonial era, therefore, left these new states generally undefended in the conventional sense (Malaysia was, perhaps, the exception) and with forces of doubtful loyalty to the ethnic group predominant in the government (for the British colonies), or forces with ties as close to the colonizer as to their own regime (for the French), or forces bereft of virtually any external loyalties whatever (for the Belgian). In a sense, the European experience had been reversed; the creation of the state had preceded the creation of the national army. Under the circumstances, it is not surprising that so many problems arose as a result.

Transition to Independence

Given the structure of the colonial forces, it was fortunate that so many of the colonies achieved independence peacefully. Although most of the borders drawn by the imperial powers made sense neither ethnically nor geographically, no new state had the forces to contest the status quo, and indeed the first act of the new Organization of African Unity in 1961 was to proclaim that the colonial borders were something not to be fought over. The internal debate in each new state over the nature of the new regime and the division of power was (except in the Belgian Congo) also initially peaceful, but this peace was not to last as long as the international one.

The British colonies in northwest Africa became the states of Ghana, Sierra Leone, and Nigeria (British Cameroon became part of Cameroon). Ghana was the first new Black African state. When it gained independence in 1957, most of its army officers and a good many sergeants were British. The plan was to replace these officers and NCOs gradually, with the transition to a fully African force scheduled to take until the mid-1960's. Ghana's first president, Kwame Nkrumah, had considerable pan-African ambitions, and the military force he had at his disposal was large by African standards (averaging over four men per 1,000 population in the middle 1960's). Moreover, because of Ghana's symbolic importance as the first African colony to achieve statehood, a great many countries besides Britain provided it with arms, advisers, or other military aid, including Canada, Italy, India, Israel, Pakistan, the Soviet Union, and the United States. This meant that, in addition to its army, Ghana came to possess among the largest and most sophisticated naval and air forces in West Africa.

In a pattern to be repeated elsewhere in former British Africa, all the remaining British officers and NCOs were dismissed in 1961, several years ahead of schedule. Another pattern was set, which was to be repeated in several new states, when Nkrumah established a special Presidential Guard Regiment, in this case extensively trained and equipped by the Soviet Union, to protect himself should his large military force become disloyal. He had reason to be concerned, because his army was more "British" in training and organization than most others, and his own autocratic leadership and leftist rhetoric were uncomfortably un-British for an army of that tone.

The Ghanaian army grew out of the Gold Coast Regiment of the RWAFF, which had had a distinct organizational identity for more than 15 years. It had the only African officer commissioned during the Second World War, plus several African officers who had attended Sandhurst and other British military schools in the 1950's. As a group, these officers were less radical than their civilian counterparts, were interested in retaining military relations with Britain, and were imbued with a British-transmitted attitude that saw armies as the custodians of true patriotism (Matthews in Keegan 1979: 828). Led by an officer corps that was probably the most strongly indoctrinated of any in Africa with the British vocational view of the officer's duty, the Ghanaian army remained out of politics longer than one might have expected, given the gap between Nkrumah's politics and those of his soldiers. In the end, though, Ghana is perhaps the clearest example of the tensions that arose in many new states between the military and civilian institutions, and the coup of 1966 was only the first of a long series of military interventions in politics that continues to the present.

Sierra Leone, granted independence in 1961, was in some respects a settler regime, dominated by the descendants of slaves freed from throughout the British Empire. In this circumstance there was no great political gap between the civilian and military leadership in the early years of independence, as there had been in Ghana. The British officers were replaced by Africans in the mid-1960's, but not before the army had gained professional experience with the U.N. forces in the Congo during 1962-63. The 1966 Ghana coup, however, provided an attractive precedent. In 1967, the ruling party invited the small army to intervene to overthrow the results of an election it had lost, and promptly found itself unable to invite them back out again. Throughout this early period, however, the armed forces remained very small and of no regional consequence.

Nigeria, which became independent in 1960, is the largest state in Africa, with a population of almost 100,000,000, and as the possessor of large oil reserves, one of the wealthiest. Nowhere else did British colonial military policy have such a catastrophic impact on the first decade of independence. Nigeria's armed forces were large compared with others in Africa (20,000 men in 1961) but small considering the size of the country, and were still largely British-led at independence, with only about 25 percent of the officers Africans. By 1966, the last British officer had left on schedule, but the army was ethnically out of balance. The Ibo tribe, a coastal Christian group, accounted for almost half the officers, whereas most of the troops and the general population were Hausa or Yoruba Moslems from the north and west. The Ibos were relatively more powerful in the officer corps than in the government at large, and having seen other coups succeed in Africa, they attempted one themselves in 1966 as part of a continuing ethnic power struggle. It resulted, however, in a bloody countercoup and pogrom in which as many as 40,000 Ibos may have died. The Ibo region, Biafra, seceded in May 1967, and in the terrible civil war that followed, millions died of wounds or starvation before the central government triumphed in 1970.

The military forces of these three states all illustrate some of the immediate problems that resulted from British colonial policy. None had African leaders at independence, and those who took charge later were placed in positions of heavy responsibility with very little experience by the standards of the developed world. Moreover, the armies were not only ethnically unbalanced, but ethnically different on the whole from the political leadership. On the other hand, the three armies also serve to illustrate the great variation in size and strength found among the military forces of new states, ranging from states like Sierra Leone, which never created a force of significant size, and Ghana, with regional pretensions if not regional capabilities, to states like Nigeria, a sleeping giant whose forces only expanded to a credible size with the coming of civil war.

East Africa was generally less economically developed than the west and had a shorter history of close contact with Europeans, since the slave trade with the Americas took place on the west coast. On the other hand, with colonies like Kenya and Rhodesia so heavily settled by Europeans, the independence movement there was a complex and sometimes violent process. Moreover, owing to the lower level of development in East Africa, the transition to independence was generally more difficult for its armies than for those of the west.

The British East African forces were organizationally separate from those in West Africa, and the two were not employed in each other's territory. Kenya had seen a widespead insurgency by the Mau Mau during the 1950's, but the Mau Mau were put down, and the post-independence army was formed from loyalist British forces. Like the armies of Uganda and Tanzania, it was formed from the KAR regiment. That colonial force had fewer African officers than the RWAFF, and its role in helping to suppress the Mau Mau insurgency made it suspect in the eyes of African political leaders. As a result, the new armies in all three countries got off to a shaky start when mutinies (not attempted coups) occurred in 1964 over pay and the continued presence of British officers. The British were invited to put down the mutinies, which they did with little fuss. Though that particularly embarrassing problem has not occurred since, it caused at least Kenya and Tanzania to place a high value on political control of the military, even though they came to take quite different paths in achieving it.

To the south, the Federation of the Rhodesias and Nyasaland had smaller-than-average colonial forces because of the Southern Rhodesian government's concern about arming Africans. This pattern was followed after independence in Malawi (the former Nyasaland), which created a very small force that relied on about 50 British military advisers and British subsidies through 1973. The conservative nature of the regime and the country's neutral stance, first regarding Portuguese policy and later Rhodesia, served to keep Malawi out of international difficulty.

Zambia (formerly Northern Rhodesia), in contrast, which sits astride the principal lines of communication with Zimbabwe and Mozambique, has been much affected by the conflicts in those countries.* Like Malawi's forces, Zambia's were small, had not been part of the KAR, and did not mutiny at independence. The new regime, fairly conservative by African standards and concerned about the revolutions on its borders, maintained close military ties with Britain; the army commander was British until 1970, and other British officers were active members of the force until 1972. This large British presence tended to keep the army in the barracks and give the impression that the military would be able to protect Zambian neutrality.

The former High Commission Territories of Lesotho, Swaziland, and Botswana, completely dominated by South Africa, could pose no military challenge to their powerful neighbor and generally did not bother to create armies at all, although the 900 Lesotho national policemen were well enough equipped to be considered an army in the African context.

Finally, to complete the tour of former British African possessions, we look to the north, where we find the only case in which irredentist designs came into play in the immediate post-independence period. When British and Italian Somaliland were combined to form Somalia in 1960, about a million people who were ethnically Somali lived outside its borders, in Djibouti, Ethiopia, and Kenya. The new government at once proclaimed its intent to annex the Somali-inhabited parts of those countries. In 1963, the regime rejected a U.S.-Italian military aid package as too small to carry out this policy (Dupuy et al. 1980: 253), whereupon the Soviet Union agreed to meet the country's military needs. In this, the first former British colony to benefit from Soviet military largess, was created what was, for the size of the population, the largest army in Africa (9.74 military personnel per 1,000 population in 1970), and by far the most heavily equipped (over 200 tanks in 1972). Somalia, however, was an exception to the rule elsewhere in former British Africa that military and civil policy concerns were internally focused following independence.

Although most French possessions achieved independence before their British neighbors, it was clearly a less independent independence, especially for the military. Arrangements for dividing French colonial forces among the new states were made as part of the "Treaties of Cooperation" that each new state except Guinea signed with France. These treaties were a package; thus a territory could not avoid a mili-

* The circumstances of Zimbabwe are so peculiar that its inclusion would have led to many data coding problems, but in any case it became independent (under Black rule) after the cutoff date for this study.

tary relationship with France if it wished to continue receiving general aid (Matthews in Keegan 1979: 828). The treaties provided for basing rights in some territories and staging rights in others, and prescribed a French monopoly with respect to military aid and cooperation. The French policy envisioned that external security would remain a French or a French Community problem, and that the individual forces would be used independently only for internal security. For most of the former French territories, this relationship in fact held until the late 1960's, with the result that French penetration of these new states was as extensive in the military domain as in the economic and political spheres.

Former French African possessions lie in two general regions: the belt of land at the southern end of the Sahara stretching from the Sudan to the Atlantic and the equatorial region generally east and south of Nigeria. The island of Madagascar in the Indian Ocean was also a French possession. The new states of the northern tier all share problems of ethnic rivalry between Berber nomads in the north and Black tribes in the south. All became independent in 1960, the easternmost being Chad. More than 500 French officers and men were assigned to Chad as advisers after independence, and the French army continued to control the northern part of the country until 1965. Nevertheless, the nation-building process never really worked in Chad. The military continually broke apart along ethnic lines, the central government never achieved control over the northern part of the country, and by the 1970's a multisided civil war was more or less in constant progress.

To some extent these problems afflicted all the former French territories in the region, but nowhere to such a disabling degree. Niger had the same ethnic tensions but not the same open rebellion as Chad, and there was a smaller continuing French military presence. Upper Volta had French advisers but no garrison. Mauritania's French garrison withdrew in 1966, but French advisers remained. The only French Community state in this region to sever its close post-independence ties to France was Mali, which was rewarded by the Soviet Union for its action with considerable military aid. This aid made Mali, on paper, the strongest nation in the region, but its military forces were still no match for the French garrisons of its neighbors.

The southern and coastal states of the French Empire were generally richer, more densely populated, and less torn by ethnic rivalries than the northern ones. All except Guinea maintained close French ties immediately following independence. When Guinea chose to leave the French Community on independence in 1958, the French withdrew all military personnel and equipment. Some 10,000 of the 22,000 Guineans in the French army chose to remain in it, and the others were discharged. With military aid from the Soviet Union, China, and Cuba, Guinea produced a military force that, like Mali's, was about the same size as others in French Africa, but better equipped. In addition, the government created a party-based paramilitary organization called the People's Militia that was as large as the army. The Guinean economy has generally performed poorly since the French withdrawal, but the elimination of the French presence had none of the harmful impact on military power that it had on economic performance.

Senegal maintained a close association with France. Dakar had been the headquarters and embarkation port for French West African forces, and the Senegalese were considered the best African troops of the colonial army. The post-independence Senegalese army, though not large, was unusually well equipped for a military force still so closely linked to France. Like Senegal, the Ivory Coast created an army of moderate size and sophistication, supported by a large French presence. Gabon created a small military force, of about average size, in proportion to its 500,000 people, for French Africa. Here too close ties with France were maintained, and when a coup was attempted in 1964, the French intervened with force to protect the government. In contrast, Togo and Benin (formerly Dahomey), where the French presence was much smaller than elsewhere in the region, saw numerous military interventions in government after independence (Togo had the first successful African coup).

Cameroon was formed from British Cameroon and French Cameroun in 1961 (the French section was independent in 1960). It is ethnically very diverse and had some insurgency problems; about 10,000 French troops remained in the country until 1964 to combat insurgents. The French also kept close ties with the Central African Republic, and a small garrison was stationed there. Those troops did not act to prevent the army chief, Jean-Bedel Bokassa (a former French captain), from seizing power in 1966, but they did intervene to protect him in 1967.

The last of the former French colonies on the continent is the People's Republic of the Congo (Congo-Brazzaville). There a leftist regime came to power in 1963, and in the pattern of Guinea and Mali, the French presence was promptly reduced. Beginning in 1963, advisers from the Soviet Union, Cuba, Algeria, and several other states were employed not only in training the Congo army, but also in training insurgent groups from surrounding African countries and Angola. Like the other states that took this route, the People's Republic of the Congo ended up with a better equipped military than any of its Francophone neighbors.

The French administered Madagascar as part of their African territory, and the few troops it drew from there were incorporated into their African units, though the population is predominantly of Malay origin. After independence the French retained a large base at Diego Suarez, along with other island garrisons numbering over 4,000 troops. Madagascar had a long history of insurrections against the French, and almost no native troops were recruited there. Perhaps because of this history, the Malagasy Republic maintained a military force quite small for a nation of 7.5 million people, and even the navy was small compared with the navies of the French West African countries. Malagasy is an interesting case because, although the army was smaller even than the national police force, it was more concentrated around the capital and thus successful in its attempts to intervene in government.

France's continuing large presence in most of its former colonies produced a certain stability, at least initially. In general, where the French maintained garrisons there were fewer military coups in the early years of independence than in either the former British possessions or those countries that rejected a continuing close relationship with France. On the other hand, the French sometimes gave at least tacit support to coup attempts, and the more radical regimes were probably quite fortunate to be spared a continuing French influence on their armies. The relatively low degree of ethnic polarization in many of the French-trained armies may also account in part for their reduced level of internal conflict. There is evidence, moreover, that a large French presence has tended to inhibit the growth of local military power. In all, the former French colonies were generally smaller and poorer than the British colonies, and their military forces, on average, were smaller (controlling for population) and less well equipped.

The only other major African power to grant independence without the provocation of a revolution was Belgium, but its former possessions had, collectively, the worst time achieving any kind of stability and prosperity as states. As noted earlier, the Belgians did very little to provide for African leadership after independence. In Zaire, the Force Publique immediately mutinied over their lack of improved status under the new regime. The military had no real allegiance to its new national leadership and could perceive no advantages under the new regime. Although independence reduced the credibility of the Belgian officers, there were no African officers to supplant them. Both the new regime and the Belgians handled the mutiny poorly, and in the spreading disorder the province of Katanga seceded, provoking a U.N. intervention that left 15,000 to 20,000 troops in occupation until 1964.

If possible, things went even worse in Burundi and Rwanda. In Burundi, the minority Tutsi tribe had dominated the Hutu majority for some 400 years, including the colonial period. Under the Belgians, who replaced the Germans under a League of Nations mandate, the centralization of government reduced a complex of conflicting geographic and ethnic loyalties until only the Tutsi-Hutu split remained (Weinstein 1976a: 8). In 1965, a Hutu attempt to seize power failed, and most Hutu intellectuals and army officers were killed. Earlier, in Rwanda, also long dominated by Tutsi, the Hutu had prevailed in a U.N.-supervised election in 1961. In the violence that followed independence in 1962, some 160,000 Tutsi fled the country in the face of an army united by ethnic Hutu loyalty and supported by the majority Hutu population.

The failure of the Belgians to prepare any national leadership in these three countries extended to military as well as civil matters. In the aftermath of independence, the military forces of each became more tribally pure and less representative of the population. None were effective agents of internal security except in the exercise of raw suppression, and none could provide any credible defense against an even moderately well-equipped external threat.

Outside of Africa the transition to independence generally proceeded smoothly. Kuwait was a British protectorate until 1961, and it maintained a close military relationship with Britain and Pakistan after independence. When the British pulled out, Iraq claimed part or all of Kuwait, but the British sent a show of force to prevent annexation. The new Kuwaiti military was mostly foreign, British and Pakistani, hired to defend the lightly populated emirate.

Malaysia survived two separate military threats to its existence, acquiring in the process an extremely powerful navy, a large (but not for its population—six men per 1,000 population in 1972) and well-equipped army, and a moderately large and well-equipped air force. The Communist insurgency of the 1950's professionalized the army, and the Indonesian assault of the 1960's cemented its ties to other British Commonwealth states in the region. Singapore broke away from the Malaysian Federation in 1965 because of ethnic hostility between the Chinese majority in Singapore and the Malay majority of the federation. It depended on the British presence for defense through 1967, taking care not to let its military procurement worry Malaysia.

The island of Mauritius, some 500 miles east of Madagascar in the Indian Ocean, was a British possession, though the French had first colonized it and French remains the language in general use. The island maintained only a small police force as a military establishment. Papua New Guinea gained independence in 1975, ending a U.N. trusteeship under the administration of Australia, with which it retained close military ties. It had no air force to speak of, a small navy of capable patrol boats, and an army several hundred strong used for internal security.

Three former British colonies in the Caribbean region are large enough to warrant inclusion in this study. These possessions—Jamaica, Trinidad-Tobago, and Guyana—were so small and remote from other concerns of the British Empire that no significant colonial forces were in place prior to independence, and the regional hegemony of the United States was such that none made any significant move to bolster its military after independence. The 1st West India Regiment was divided between Jamaica and Trinidad-Tobago, each ending up with a 1,000-man battalion. Guyana has fewer people than either island state but maintained a slightly larger army. Although lightly equipped, it was employed several times (mostly as a show of force) in disputes with Surinam and Venezuela over the ill-defined borders in the interior. But like the others, it remained largely out of domestic politics. The former Dutch territory of Surinam likewise had only a small force of soldiers when it gained independence in 1975. Unfortunately, Surinam's small civil elite was decimated after failing to prevent a coup led by an army sergeant.

Summarizing across such a great variety of new states is difficult, but some generalizations are possible about the ones we have touched on, all of which achieved independence peacefully. The capabilities of the new military forces varied according to the level of the former colonizer's continued presence. This was the case with Britain and Singapore as well as very obviously the case in the former French territories. Those new states with a large British or French military presence in the early years had smaller and less well-equipped forces than those that rejected their former rulers.

The metropole's continued presence also smoothed the transition to independence for the military. The transition was difficult where pre-independence African leadership was most limited, and the pullout after independence most abrupt. Where mutinies occurred and the metropole was militarily powerful enough to reimpose order (British

East Africa), a breathing space was granted for military integration into the independent regimes. Where the metropole was not sufficiently powerful (Belgium), the result was chaos.

In general, the transition to independence carried with it the seeds for the patterns of future development. With the exception of Malaysia, all the new states began their existence with armed forces that were inadequately equipped and deployed for national defense and, further, had no experience in thinking about or planning for that defense. Many were ethnically out of balance with the rest of the country and were led by men whose education and training had been substantially different from that of their civilian counterparts. The stage was set for the significant expansion of these new forces, not only in size, but also in their domestic political power, during the next quarter-century.

Revolution and Its Aftermath

What differences, if any, mark the military's transition to independence in the new states that did not achieve statehood without a struggle? To begin with, as one would guess, the new military forces of the revolutionary regimes, unlike those of the other states, were not formed from the colonial forces, did not develop independently of the new civilian rulers, and maintained an uneasy relationship at best with the former metropole in the years following independence. Nevertheless, these factors did not have as great an impact as might have been expected because the revolutionary armies were quite different from those armies that helped to create the present states of Europe or the Western Hemisphere. They did not, for example, actually win the military struggle for the territory of their new countries, but rather exhausted either the patience or the treasuries of their colonial enemies. They were not, therefore, capable at independence of handling internal or external security assignments markedly better than their nonrevolutionary contemporaries, and though the military and civilian rulers of these new states shared long years of anticolonial struggle, the two groups developed some of the same tensions that appeared in other new states.

The largest and most costly war of independence took place in Algeria, the one French territory in Africa that had a large settler population. It demonstrates as well as any case the limits to the role revolution played in post-independence processes. The Algerian Revolution lasted from 1954 to 1962, and probably cost a million Algerian lives.

As in other African anticolonial wars, the rebels were unable to defeat their enemy militarily, so that the French managed to retain control over most of the territory. The cost of their brutal counterinsurgency campaign was high, however, and as many as 500,000 French troops were tied down in Algeria protecting the million French settlers there. The cost, in the end, was too high, and De Gaulle agreed to a referendum that led to independence in 1962.

The rebel Armée de Liberation Nationale (ALN) was under the political control of the Algerian government-in-exile in Cairo. By 1962, it mustered about 150,000 troops, but most of these men were in its external wing, stationed in Tunisia or Morocco and not involved in fighting the French. The guerrilla war was prosecuted largely by a varying number of ALN irregulars stationed inside Algeria. Although individual men were frequently transferred between the ALN's internal and external wings, there was a serious rivalry between the leaders of the two groups.

The rivalry came to a head at independence when the political leaders of the revolution (under Ahmed Ben Bella), which had signed the treaty with the French, and the military leaders of the external forces (under Houari Boumedienne) sought the allegiance of the commanders of the internal forces. When some of these commanders were unwilling to surrender control of their territories to groups that they felt had not won the victory, Boumedienne deployed his numerically superior external forces against them. In the fall of 1962, he managed to secure an agreement effectively recognizing the supremacy of his faction (Dyer in Keegan 1979: 13).

Once the Algerian government was organized, Boumedienne was made defense minister and commander of the armed forces. He promptly purged nearly all the officers and discharged most of the men from the internal wing of the now unified army. Only 10,000 of the 60,000 men in the new Algerian army came from the internal forces. In 1965, when Ben Bella tried to rein in the power of his defense minister, Boumedienne staged a military coup, and the army became the sole effective political power in the country.

Algeria illustrates what revolution may imply for the transition to independence. The political and military wings of a revolutionary group are not necessarily united in philosophy or strategy, nor are they free of personal rivalries. Guerrilla war is by definition loosely controlled and depends on the initiative and independence of local commanders. Unifying the country after independence is not a trivial task in such circumstances, though no one was more brutally successful at it than Boumedienne. He became the country's leader because he was able to gain complete control of the military, and given no other institutions as well organized or developed in competition, with the military went all the political power.

Like its nonrevolutionary counterparts, however, Algeria began independence with a small, nonprofessional military. Well equipped by the Soviet Union in its early years, it was, nevertheless, not the force of veteran Berbers who harried the French, and its performance in early border skirmishes with the Moroccans was not impressive. It was by no means established at the outset how effective and stable an institution of national (if not so far international) power it would become.

Portugal may seem an odd country to have kept hold of an empire into the 1970's, especially one whose population was twice as large as its own, but it is a measure of the low levels of political and economic development in its possessions that Portugal was able to manage the task, giving up only when the military, disenchanted with colonial wars, overthrew the civilian dictatorship in Lisbon in 1974.

Angola, in 1961 the site of the first open revolt against Portuguese rule, was not the largest of Portugal's colonies (population about 6,000,000), but it was the one in which the most costly conflict took place. Unlike the other major Portuguese African possessions, Angola did not have a single, unified revolutionary political and military organization. Four groups in fact were active there, organized along tribal and geographic lines. That four separate organizations could flourish at once was due as much to the availability of outside sponsors as any internal factors (the Soviet Union, the People's Republic of China, Zambia, Zaire, and later South Africa), but the problem for Angola was that, given each group's territorial autonomy and ethnic orientation, no single one could manage to dominate or direct the struggle for independence.

In contrast to the French departure from Algeria, the Portuguese left Angola without signing an agreement with any faction; they simply packed up, vacated their offices, boarded ships, and left in November 1975. The Popular Movement for the Liberation of Angola (MPLA), the Soviet-favored group, had drawn support from the Ovimbunda tribe as well as the mixed-race community in the capital, Luanda. It controlled the capital and the immediate surrounding area. Since Luanda is a port city, it was fairly easy for the Soviet Union to funnel in military supplies to the MPLA. It also gave access to the 14,000 Cuban troops who joined in the MPLA's fight against its revolutionary competitors and an invading force from South Africa. The South Africans were not defeated in battle, but lacking American support, they saw themselves overmatched in the long run and withdrew. By early 1976 the MPLA clearly had the upper hand in the civil war, and it gained wide diplomatic recognition as the new government of Angola.

Despite Soviet and Cuban assistance, however, the MPLA was unable to retain control over large areas of Angola, notably those regions inhabited by the Bakongo people, who supported the Zaire-based, Chinese-supported National Front for the Liberation of Angola (FNLA), and a tract in the southeast, where with South African support the National Union for the Total Independence of Angola (UNITA) was able to establish itself as a virtual state within a state. Insurgency also continued in the enclave of Cabinda, a region north of the Congo river delta that is small but important because of its oil production.

Control over the MPLA itself was also contested shortly after independence when the interior minister and a number of army officers attempted a coup. Although the coup was suppressed, it was at the cost of increased representation for the army in the political and central committees of the MPLA Congress. The troubles of Angola are a clear measure of how far from effective national armies the guerrilla forces that gained independence were, and how fortunate most new states were to avoid the kind of "help" that so many neighbors and distant powers gave to the country's contending revolutionary forces.

In both large Mozambique (population about 9,000,000) and much smaller Guinea-Bissau (population about 600,000), the transition to independence was made much smoother by the fact that only one group dominated the revolutionary struggle. Nevertheless, in neither case did that group have an easy time asserting control in the vacuum left by the departing Portuguese. In Guinea-Bissau, the People's Revolutionary Armed Force (FARP) was able to organize in exile in Guinea, much as the ANL had done in Egypt during the Algerian Revolution. It was supported by the Nalu and Balanta peoples of the territory but not by the Fula, who largely remained loyal to the Portuguese. Although there has been none of the widespread guerilla warfare seen in the two southern states since independence, there have been exile invasions, problems with Fula unrest, and tensions between the party and the military leadership reminiscent of Algeria.

Mozambique became independent in 1975 after a 12-year insurgency. The Frelimo party, which organized nearly all the resistance, ran the government after independence. As in the other territories, however, the Frelimo had not won a military victory over the Portuguese, and had serious problems asserting control in the face of Rhodesian and South African incursions and the presence of armed African veterans of the Portuguese colonial forces. It also faced the same problems the others did when nearly all the Portuguese settlers left, depriving the territory of most of its technical workers and nearly all the people with business and managerial experience. Nevertheless, in the face of a major external threat, the Frelimo had considerable success in maintaining internal stability and in integrating the military into the party's political institutions.

Algeria and the Portuguese colonies represent the only cases of classic anticolonial revolutions in the period of this study, and yet on examination they are less "classic" than they seem. In no case was independence won by a united revolutionary politico-military organization that defeated the metropole and took over the territory. The similarities to the regimes that had peaceful transitions are as striking as the differences. The initial military forces were small and not adequate for external security, or even for internal security in every case. There were tensions between the civil and military leaders, and tensions within factions of each. The revolutionary path to independence did not necessarily lead to an initial military situation markedly different from the peaceful route.

Three of our cases involve former British colonies that underwent revolutionary struggles, but none was a textbook case of anticolonial revolution. It is difficult to label what happened in Aden because the transition to independence in 1967 as the People's Democratic Republic of Yemen (South Yemen) was not very peaceful and did not proceed as the British had planned. Nevertheless, the British did officially recognize one independence group before they left and provided some initial military equipment and organizational support. That group, the National Liberation Front, was able to eliminate most of the effective resistance by the end of 1968, during which it signed an arms aid agreement with the Soviet Union. South Yemen had border disputes with both North Yemen and Saudi Arabia and has been used as a base by insurgents from Oman. Although political control of the regime changed hands several times after independence, the military organization first recognized by the British has remained in place.

In Cyprus, as in Aden, the British left under duress and the armed forces were not formed from colonial units. Independence was achieved in 1960 through a compromise between the Greek faction, which had fought the British to achieve union with Greece, and the Turkish minority, which preferred either continued colonial status or partition. The military forces of the new state were split 60–40 along ethnic lines and were far too concerned with maintaining order between the Turkish and Greek communities ever to develop into a national force oriented toward defense of the island's sovereignty or any other goal.

Although Bangladesh can properly be called a former British colony, the British had been gone for 24 years, and the country gained its independence in 1971 in a war of secession from Pakistan. Since British recruitment policy in India had favored Punjabis and Pathans, when the British left in 1957 the Pakistani army was nearly exclusively manned by officers and men from West Pakistan. During the next 24 years a small percentage (usually less than a tenth) of the Pakistani army was recruited from Bengalis in East Pakistan (even though they constituted a majority of the population of the united country), but Bengalis were not given positions of authority and were not imbued with the same traditions as the older regiments from West Pakistan, which had long histories as elements of the British Indian Army.

The army intervened in Pakistani politics several times in the next two decades. But then, in 1970, Gen. Yahya Khan, who had taken over the presidency the year before, promulgated a new constitution that provided for proportional representation in the National Assembly. In the elections that followed in December, the Awami League of East Pakistan captured 160 of 162 seats available to it in the assembly, gaining a majority over Ali Bhutto's Pakistan People's Party. Yahya Khan thereupon refused to convene the assembly, and after two months of negotiating, the Awami League proclaimed itself in power in the east in March 1971. At this time most of the Bengalis in the Pakistan army were stationed in the west; they were quickly disarmed and interned. The 80,000 non-Bengali Pakistani soldiers in the east were able to seize and hold all the major population and communication centers, but could not maintain order against the guerrilla uprisings that spread across the country. The army responded with a horrific campaign of suppression in which as many as 1.5 million people may have been killed, including nearly every member of the educated Bengali elite. The Indians finally intervened in December 1971, and after 13 days of fighting compelled the surrender of all Pakistani forces in the east. In the first months of 1972, the independence of Bangladesh was recognized by nearly every state in the world.

Independence by no means ended the problems of Bangladesh in any sense, including a military one. With 60 percent of the police force dead and independent guerrilla forces controlling large territories, attempts were made to integrate the guerrilla forces (the Mukti Bahini) into an army organized of those regulars who had avoided capture in the east at the start of the revolution. Although this was accomplished on paper, the new force was shaken by discord. Most of the regular Bengali officers and men had spent the struggle for independence and the first year afterward interned in West Pakistan. Having missed the fighting and the hardships of the struggle, they returned from captivity in 1975 to take lower positions than they had once held. Discord over military prerogatives reinforced disagreements over political strategy as various revolutionary factions attempted to organize the officers and men to their own ends. As a result, Bangladesh suffered a series of often-bloody coups and attempted coups in the first years of independence, and no consensus on the role or structure of the military developed either within the military or outside it.

Whether peaceful or revolutionary, in sum, the independence process left recent new states in situations that would not have been possible in most other periods—with military forces too weak to defend them, but often strong enough to control the government. With these initial conditions in mind, we are now ready to begin looking with rigor at the post-independence period and to measure the changing military capability of these new states.

THREE

Acquisition Patterns

Military capability can be measured in any number of ways, but most of the obvious indexing techniques are inappropriate for the task at hand. In the simplest sense, military capability is what enables a state to win a war, and the most straightforward way to measure it is to rank a set of armed forces according to how well they have done in combat against one another. Even if there has not been enough warfare to enable the calculation of a unique rank ordering, such combat as has taken place may at least provide standards of force structure against which all states can then be judged. Our concern, however, is to examine the quest for military power as a driving force in the development process, not to explain the results of interstate competition or predict the outcome of such competition, and so, a rank ordering along these lines is not useful. For that matter, there has been so little interstate war among the states of our set that a meaningful empirical scaling based on war results is impossible.

Likewise, broader scales of national power are not appropriate to the project at hand. To compare an internal national process, the creation of military capability, across a very wide set of states and a long period of time requires indicators derived from data that can be collected at the state level and employed without reference to interstate conflict.

Since the concern *is* with an internal national process, we must clearly devise an index that measures how important military power is to the governments of new states. Although examining defense costs as an economic burden might seem a good way to do this, it would ignore not only outside assistance, but also the efficiency with which the available resources were used. At the core, what is needed is an index that shows how serious the governments of new states are about armed power. It must account not just for how much is spent, but for how well it is spent. It must account not only for how much is bought, but for what is bought. If, for instance, a government spends a great deal on plush transport aircraft for its general officers, the purchase might have the same economic impact as an investment in tanks, but it would indicate that employable military power was of less concern there than in a state that bought the tanks. A country can be very serious about military capability even if it gets every bullet free of charge from a patron, and the index needs to reflect that fact. But whatever the measure chosen, it needs to have a qualitative component as well as a quantitative one.

Finally, with the wide variation among new states that we saw in Chapter 2, the difficulties of defining standards for measuring military capability are compounded. In geographic terms alone, there is so much variation across the sample that an argument could be made for creating a scale for each state or region so as to consider tanks more important to Upper Volta, for instance, and less important to Malagasy. This solution is clearly unworkable, however, because placing 46 states on 46 different scales is the same as having no scale at all, and would at best provide no more than a comparison of each government's view of its own defense needs with that of the coder. What we need, in sum, is a comparative scale that will measure how much attention, in both a quantitative and a qualitative sense, states pay to the acquisition of military power. It need not account for or predict the outcome of interstate war, but it has to be employable in cross-national and cross-temporal analyses over a large number of countries and years. These objectives guide the choice of indicators for military capability, the collection of data on them, and their manipulation in the analysis.

Indicators

Indicators of military capability fall broadly into two categories, which may be conceived as inputs and outputs. The inputs are defense expenditures, the defense budget of a state being the input to the creation of its military capability. The outputs are weapons stocks and military personnel. This is a different use for expenditure data from the economic burden concepts discussed in Chapter 1.

Expenditure totals have often been favored in research involving military capability, and they have several important advantages over output indicators. If all appropriate items are included, then the whole range of distinct weapons systems possessed by a state are accounted for on a single scale. The scale also provides a continuity from one year to the next that may not be possible with other indicators. A new warship, for instance, may encompass technological advances that make comparisons with its predecessors difficult, but budget figures can account, in some sense, for such advances through the increased costs of more modern systems. Budget figures are also in units that can quite readily be compared with other national economic data to yield information on the costs, impact, and growth potential of defense programs.

But expenditure data also have their disadvantages. They may quite inaccurately reflect the tempo of national efforts by missing shifts in emphasis between production and research and development (Gray 1971) or between one very expensive weapons category (like tanks) and a cheaper but equally important category (like anti-tank missiles). Procurement priorities can easily affect the budget in a way that does not reflect the progress of capability acquisition, and in addition, replacement cycles can cloud all but the longest-term trends (ten years or more), especially when budget authorizations for long-term deliveries show up in single year's totals.

Nations also keep their accounts in all sorts of ways (and often change those ways), to the point where the defense budget means something different in virtually every state for which it is reported. This makes comparisons problematic, particularly between centrally controlled economies and others. In attempting to deal with this issue, some researchers (e.g., the Central Intelligence Agency working on the USSR defense budget) have attempted to translate known weapons-stock and military personnel totals into their equivalent cost in the United States or some other Western industrial country. The unfortunate result of this approach is that every time U.S. military personnel get a pay raise it makes it look like Soviet expenditures have gone up. The approach is simply inadequate for most cross-national and cross-temporal research.

Differential inflation rates also make cross-national and cross-temporal studies difficult and greatly complicate the interpretation of any results. Differences result if one chooses to inflate in the local currencies (to get a constant value unit across years) and then transform into dollars (to get a value comparable across countries), or to first transform into dollars and then inflate at the U.S. rate. Worse, exchange rates vary over time and with markets, and many arms transfers involve exchanges of currency or goods set at secret prices that are considerably lower than world market values.

Finally, expenditure figures do not account very well for arms transfers. National accounts on such matters are often secret, and those that are available often understate the value of the transfers, since much of the equipment is given away or sold as surplus at very low prices. The Stockholm International Peace Research Institute's way of approaching this problem is to assign a monetary value to each piece of hardware on the basis of such factors as price, production date, depreciation, weight, speed, and role (SIPRI 1975). But though this solves one problem of comparability, the resulting numbers no longer provide even the vaguest representation of either donor or recipient costs. The method sacrifices the directness of hardware data without achieving any of the advantages of expenditure figures for economic analysis.

The weakness of expenditure data on arms transfers is especially damaging when the focus is on states that cannot produce weapons themselves, for not only must the arms be imported—except for some small naval craft produced in Singapore, Gabon, and the Ivory Coast, *all* major weapons came from outside this set of states during the period of the study—but often the personnel to operate them as well. As noted in Chapter 2, Pakistani advisers help run Kuwait's army, the United States has kept the Zaire air force flying, Cuban advisers are or have been central to the operations of the forces of South Yemen, Ethiopia, Angola, and others, and the French have garrisoned thousands of troops in several African states. Yet nowhere in the arms-transfers figures do the costs of such foreign personnel appear, even though they have played a significant role in helping certain states increase their military capability.

For all these reasons, expenditure figures are ill-suited as indicators of military capability in this research. The problem of defining a monetary unit useful across the sample over time is next to impossible. In any case, the theoretical questions addressed here require indicators that go beyond the cost to a government (or its patron) of its defense establishment and are more directly related to the resulting capability.

Still, in choosing outputs as indicators, the fact that they have their own weaknesses cannot be ignored. Data on weapons are often classified and almost never account for the percentage, not only in stock, but actually operational at any given time. Data on force readiness would be one of the best possible guides to the seriousness with which a government views military capability. But even in the most developed states, such data are difficult to get and questionable even then. Arguments about the readiness of U.S. forces routinely surface during Congressional hearings, and doubts about the reliability of the (generally secret) numbers have been voiced from many quarters. Readiness data are simply beyond the reach of anyone doing unclassified research, and reliable readiness data are probably beyond reach, period.

Another problem with hardware data is that military power as such embraces a number of components, even in states with relatively simple defense needs. Manpower, tanks, aircraft, and gunboats all contribute their share, and no intuitively obvious formula for their combination presents itself. Even where attempts have been made to rank the utility of weapons systems on a multidimensional scale (e.g. Kemp 1970; Laurance & Sherwin in Ra'anan et al. 1978), no one has managed to combine and rank them into a single military capability scale. Nevertheless, for the purposes of this investigation, output indicators are the preferred alternative. The quantitative and qualitative results of a state's military growth can be more closely measured by these indicators than by inputs, and because predictions of war outcome are not required in the needed index, the lack of a universal formula for translating each country's weapons mix into its predicted outcome in combat will not cripple the index's utility.

Military personnel totals are the most frequently used output indicators of military capability, and their use has several advantages. Soldiers have been around a long time and are not subject to radical technological change, so they are particularly useful in a cross-temporal study. Soldiers are also fairly easily matched up across countries, a Nigerian soldier being generally much more the equivalent of a Malaysian soldier than a Nigerian ship, for instance, is the equivalent of a Malaysian one (although Nigerian and Malaysian soldiers might choose to disagree on this point). Finally, since manpower totals are not normally kept secret, the data are comparatively easy to come by.

Manpower totals alone, however, are insufficient as indicators of military capability. By themselves, they provide little measure of qualitative attention to capability, though if one could measure the level of training or the level of readiness, then manpower in combination with one of those items might do. Unfortunately, as we have seen, such data are wholly unreliable, so that this approach is unfeasible. But simple manpower totals would still be qualitatively useful if they were highly correlated with weapons-stock totals. This is not the case, however. In the forces of states varying so widely in size, sophistication, and geographical circumstance, the number of military personnel in a given force is not necessarily proportional to the quantity of its weapons stocks.

Therefore, any index of military capability that meets the needs of the study requires data on hardware as well as personnel, and will entail assumptions about that hardware as a qualitative as well as a quantitative measure. Since no weapon of modern times (perhaps not even a nuclear weapon) is suitable for all targets or invulnerable to all countermeasures, a mixture of weapons is required for a military force to be effective. Therefore, one way to check for quality in a force is to examine how balanced it is across all categories of weapons. States that devote great efforts to acquiring the most advanced weapons in one category at the expense of others may be paying great attention to one quantitative matter, but they are ignoring a qualitative aspect of military capability. A good index must reflect such an imbalance when producing a military capability score.

Of course, incorporating the concept of balance cannot be accomplished as a mindless mathematical exercise. Finesse is still required, given the variety of states under study. For example, one would not expect naval weapons categories to be filled in landlocked states, yet the scores of those states should not suffer for it. One also has to be concerned with improvements over time within categories. Fortunately, however, as it turns out, qualitative improvements within weapons categories have been more limited for this group over the past 25 years than for many more developed states, so that there are few difficult choices about how to treat improved systems within single categories of weapons.

Balance is not the only possible qualitative indicator; as mentioned above, some measure of competence or level of training and readiness is a theoretically possible (though not practical) alternative, but using balance among categories in combination with force size permits the weapons-stock figures to "speak for themselves" and thereby provide both the quantitative and qualitative dimensions demanded of the index.

Data Collection

Collecting data on the military forces of the states in the sample was a relatively straightforward task. Most of the countries were not interested in keeping their stockpile figures secret, and the period under investigation was far enough in the past that mistakes in early reported figures were fairly obvious and correctable. The greatest difficulties arose from the fact that the bulk of the arsenals in question were so insignificant in global terms that the most reliable published sources paid almost no attention to them.

The 46 states in this study have a total of 859 country-years of independence for the period 1957-81 (the average period of independence was 18.7 years). For each state I collected annual data on the number of military personnel on active duty and on every major weapon in its arsenal from the time it became independent. Except for small arms, which correlate closely with military personnel, and crew-served fieldpieces, on which data were usually unavailable or unreliable, I attempted to account for each and every type of weapon—aircraft, naval vessel, armored infantry vehicle, and guided missile system—present in each year in each state. (For a discussion of the sources for these data, see the Bibliographical Note, pp. 130-34.)

Before 1972, only occasional listings of stock of non-naval weapons were published for the majority of these states. But in general annual reports on arms transfers were available, and I used these to compute total stocks of army and air force weapons, correcting the figures with those available (non-annual) published reports of weapons actually in service. The corrections involved were minor and infrequent; they resulted generally from the crash of aircraft or the accidental loss of infantry vehicles that reduced the stock totals below what the transfer figures would have indicated. There are excellent annual compendia on world navies for the entire period, making the above procedure unnecessary for naval arms.

The collection effort created a data set containing annual totals for the 46 countries on military personnel and 325 different weapons systems. The resulting data matrix contains nearly 300,000 entries, making computer analysis essential; but as the description of that analysis begins, it is important to keep in mind what the numbers stand for, because each of the numbers in this matrix refers to a real weapon, with real costs in its acquisition, and real effects in its use.

System-Wide Growth

The best place to begin an orderly examination of the military capability data is with a system-wide look at the weapons-stock totals to see if any obvious features or patterns are discernible and to get a general picture of the data before any mathematical transformations (and the assumptions that must accompany them) are made. The results of such a look are startling. As a group, the new states display a remarkably uniform growth rate in their weapons stockpiles over the entire span of the decolonization period. Moreover, this uniformity holds right across all major classes of weapons.

Figures 3.1-3.3 plot the total number of naval vessels, land weapons, and aircraft held by all the states in the study from 1960 onward. Note

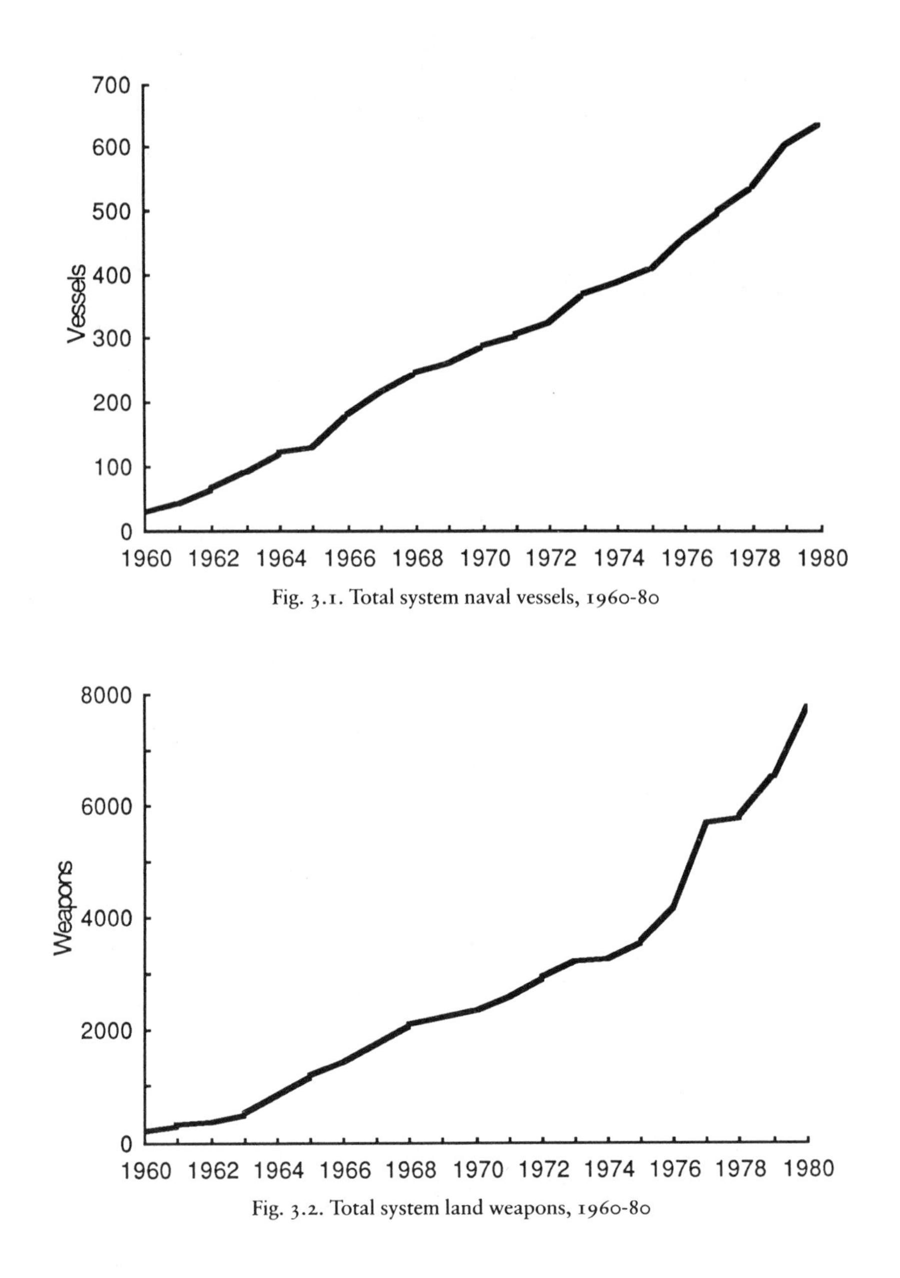

Fig. 3.1. Total system naval vessels, 1960-80

Fig. 3.2. Total system land weapons, 1960-80

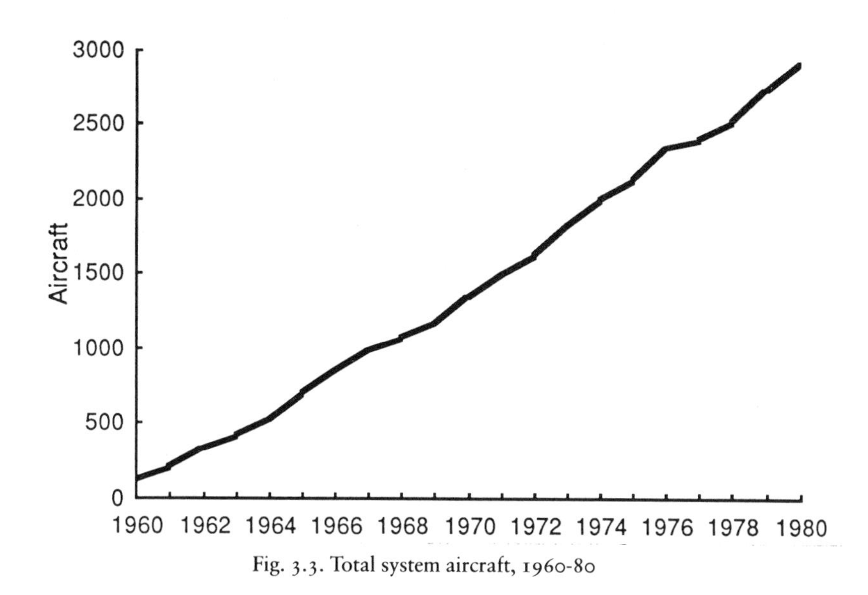

Fig. 3.3. Total system aircraft, 1960-80

that the three curves are virtually identical, and that all are nearly straight lines. In fact, the growth in naval and air weapons correlates with time at greater than r = .99,* and even the land weapons correlate at .96. As a group, the growth rates correlate with time at over .98, and as we can see, they show no tendency to depart from the pattern as time passes.

The linearity of the relationships is remarkable and, given the conventional wisdom that the rate of arms transfers to the developing world has been rapidly accelerating, quite unexpected. Remember, these are raw data being plotted, with no transformations done or outliers removed to make a pattern clearer. It is especially surprising to find the linearity holding up across the three different classes of weapons, indicating that acquisitions in a single class are not so great as to submerge variation in the growth rate of the other weapons systems.

Since there is virtually no production of major weapons in these less-developed states, Figs. 3.1-3.3 actually reflect the arms-transfer rate from outside by unit of weapons (not cost—these are weapons-stock figures). The data indicate, at a minimum, no obvious policy decisions or other time-specific actions by the suppliers that affected the rate of

* Pearson Product Moment Correlation (r).

arms transfers during the period. It appears that the system-level stockpiling rate is not driven by forces external to the system, and the system-level growth of military personnel supports this conclusion. Figure 3.4 plots the total number of military personnel over the same period. There too we see a nearly linear relationship, with the total system military personnel correlating with time at close to a .96 rate. Growth accelerates slightly during the Nigerian civil war and with the conflict over the western Sahara, but is generally uniform through 1976. It levels off thereafter; the drop after 1977 reflects a partial demobilization in Bangladesh (whose army is large enough to significantly affect system totals) and the collapse of Ugandan forces after the Tanzanian invasion. Exclusion of those cases would leave a very moderate positive growth rate for the system between 1960 and 1980.

Since military personnel are rarely imported, the linearity we see in Figure 3.4 is an important indication that a mechanism internal to the system is driving the growth of military forces. One need not look far for at least one candidate for this mechanism. Figure 3.5 shows a simple sum of all weapons (land, sea, and air) plotted together with the growth of total system gross national product. Both lines are nearly straight; in fact, the two totals correlate at greater than .97 and if plotted against each other would show a single straight line. The lines are parallel in Figure 3.5 because the vertical axes have been drawn to

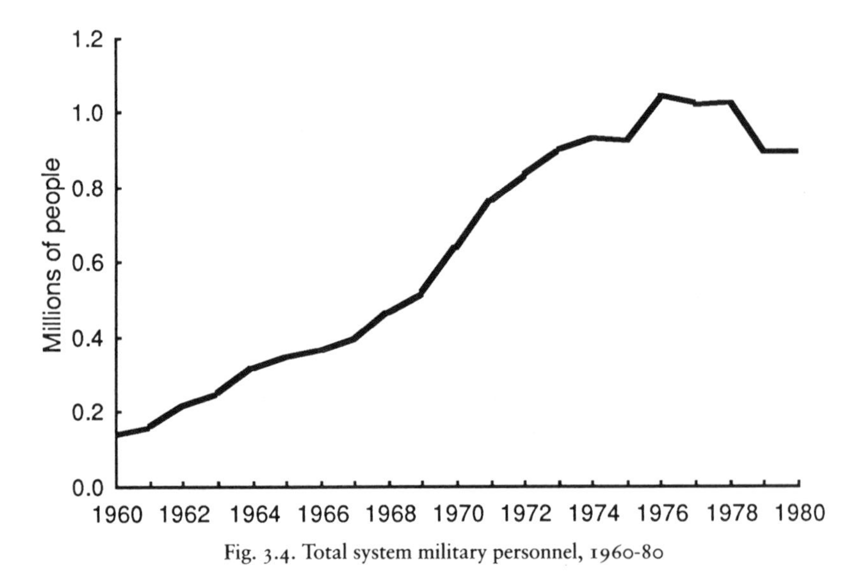

Fig. 3.4. Total system military personnel, 1960-80

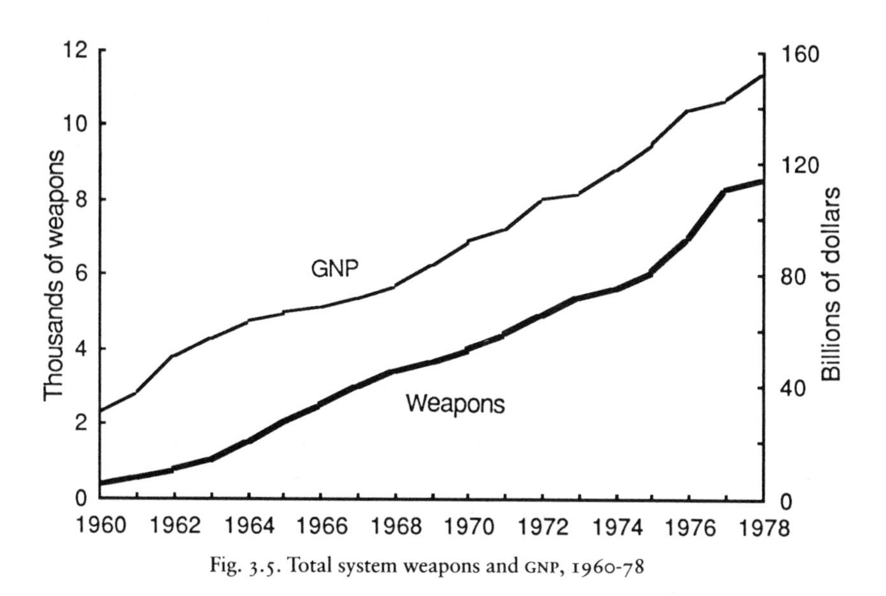

Fig. 3.5. Total system weapons and GNP, 1960-78

make them so, but the point is that both lines are straight, and plotting them together against time shows that the relationship has been consistent all across the period. Clearly, one explanation for the growth of weapons stocks is the economic growth that supports it.

It is important not to overemphasize the significance of these aggregate totals, however, no matter how startling their linearity. Aggregate stockpiles do not measure aggregate capability, and summing all aircraft as though they were equivalent (and doing the same for ship and land weapons) tells us nothing about the structure of individual forces, the behavior of individual states, the relationships between arms suppliers and recipients, or the relationships between a state's arms acquisitions and its economic performance. The graphs do not even show the totals for a system of consistent size, because the number of states grows across the period of study as additional countries achieve independence.

Nevertheless, Figures 3.1-3.5 describe some remarkable system-level behavior that will require integration into the final analysis. At least for the newly independent states, the rate of arms stockpiling remained absolutely constant for at least 20 years. Although costs certainly rose over the period due to both inflation and the growing sophistication of weapons, the unit transfer rate did not change at all. Not only was there no accelerated growth in the implements of military capability toward the end of the period, the growth of military personnel actually slowed down. The linearity in these patterns matches that of system GNP, suggesting that the growth in stockpiles is regulated by the absorptive capacity of these new states, rather than by the policies of outside suppliers.

Weapons Categories

As discussed earlier, military capability is conceived in this study as having both a quantitative component, measured by the size of a national weapons arsenal and military manpower total, and a qualitative component, measured by the balance of the national stockpile across all categories of weapons. There are a great many ways to categorize weapons, but the most appropriate for this study is by mission, with a secondary consideration being some reference to capability. That is, some missions are more demanding than others, but are similar enough that an ability to fulfill the more demanding ones implies an ability to fulfill the less demanding ones as well. For example, tanks can do what armored cars can do and also a lot more. Since buying only high-capability systems is not an efficient way to cover all missions, even wealthy states normally buy some lower-capability systems for their less demanding requirements.

In the final design, weapons were divided into 13 categories (plus a category for military personnel). Four of the categories involved naval vessels, weapons in which the 11 landlocked states in the study would be unlikely to invest. These categories are treated separately later in the index-construction process. Certain familiar categories will not be found in the classification because none of the new states possess such weapons—for example, underway replenishment ships, which are a substantial fraction of the U.S., Soviet, French, and British navies. The same obviously applies to nuclear weapons. Full details on the coding rules and a complete list of the individual weapons systems assigned to each category will be found in Appendix A. Here a brief description of the categories will suffice.

1. *Military personnel.* Active-duty armed forces (not reserves), including paramilitary forces where those forces resemble regular units in their organization, equipment, training, or mission.

2. *Utility aircraft.* Light fixed-wing and helicopter aircraft used in light transport, vertical assault, artillery spotting, or other support roles (e.g. Piper Cub).

3. *Transport aircraft.* Medium- and heavy-lift aircraft capable of hauling troops and cargo over long distances (e.g. C130).
4. *Attack aircraft.* Low- to medium-capacity aircraft usually employed to deliver air-to-ground ordnance with only limited air-to-air capability (e.g. BAC 167 Strikemaster).
5. *Fighter aircraft.* High-capability aircraft generally designed for air-to-air combat but almost always with air-to-ground capability as well (e.g. Mirage III).
6. *Trainer aircraft.* Aircraft used principally for instruction although generally capable of limited light attack roles; necessary to maintain pilot proficiency even when initial training was accomplished by a foreign air force (e.g. T28 Trojan).
7. *Armored cars.* Protected transport and attack vehicles for use in counterinsurgency and infantry support missions (e.g. Saladin).
8. *Armored personnel carriers.* Armored troop transport vehicles carrying more personnel than armored cars with generally greater protection (e.g. BRDM).
9. *Tanks.* Armored tracked vehicles mounting a large gun (71mm or larger) that can be aimed without altering the vehicle's heading (e.g. T-62).
10. *Missiles.* Ground-launched guided missiles with either anti-air or anti-armor missions (e.g. TOW).
11. *Guard boats.* Small craft having very light weapons and only an inshore patrol capability (e.g. Ham class minesweeper).
12. *Patrol boats.* Craft of either very high speed or medium size carrying weapons with at least some anti-ship capability that permit offshore patrol roles (e.g. Shanghai II PT boat).
13. *Corvettes.* Ocean-going vessels with major anti-ship, anti-submarine, or anti-air capabilities (e.g. 850-ton Dutch-built corvette).
14. *Amphibious craft.* Vessels designed to transport combat troops and supplies to shore without the need of docks or lighters (e.g. LST).

Balance and Growth

With the weapons categorization complete, we can now turn to the question of how balanced the very linear system-level growth in total weapons stockpiling was. Table 3.1 shows the correlations between military personnel, air weapons, and ground weapons; naval weapons were not included, since some countries have none. It is interesting to note that though all the correlations are positive, they are not especially high. The correlation of military personnel with other weapons

TABLE 3.1

Total System Correlation Matrix: Military Personnel and Land and Air Weapons

Category	Utility aircraft	Transport aircraft	Attack aircraft	Fighter aircraft	Trainer aircraft	Armored cars	Armored personnel carriers	Tanks	Missiles	Personnel
Utility aircraft	1.0000									
Transport aircraft	0.5977	1.0000								
Attack aircraft	0.1437	0.2384	1.0000							
Fighter aircraft	0.3536	0.5751	0.3155	1.0000						
Trainer aircraft	0.6445	0.6459	0.3283	0.4286	1.0000					
Armored cars	0.4611	0.5609	0.3996	0.2348	0.5475	1.0000				
APCs	0.2806	0.4546	0.2388	0.8324	0.4058	0.0739	1.0000			
Tanks	0.2481	0.5278	0.2041	0.8857	0.3732	0.1093	0.8666	1.0000		
Missiles	0.2758	0.2929	0.2147	0.5213	0.3980	0.3359	0.4814	0.5096	1.0000	
Personnel	0.5449	0.4659	0.0491	0.3815	0.4469	0.3781	0.2638	0.2952	0.1645	1.0000

systems, for instance, nowhere reaches 0.6, and for three of the four categories of ground weapons (armored personnel carriers, tanks, and missiles) it is below 0.3. Clearly the point made earlier that manpower totals were unlikely to correlate highly with most weapons-stock figures was correct. In fact, the categorization scheme seems to have tapped a significant source of variation in weapons acquisition, at least at the system level, as shown by the differences across categories. Table 3.2 gives the correlations for the naval vessels held by the 35 states that are not landlocked. As before, the correlations are generally not very high; in fact, they have an even lower average value than the other weapons have (.45 versus .58).

The picture is not so chaotic, however, as a first glance would indicate. Note that such high-capability systems as tanks and fighters correlate more strongly with each other (.8857) than with lower-capability weapons such as armored cars and utility aircraft. The same holds true for low-capability systems; armored cars and utility aircraft, for instance, are correlated more highly with each other (.4611) than either is with fighters or tanks. This indicates that there is some order in the structuring of forces displayed at the system level, with states apparently attempting to maintain a similar level of capability across a broad spectrum of categories rather than concentrating on just one area of warfare. A cross-temporal examination of the intercorrelation of weapons provides further insights.

Figure 3.6 shows the mean system-wide correlations for military personnel and non-naval weapons for the years 1958–81. The scores are initially quite low, and for the first three years the curve appears erratic. This is due to the small number of states in the system, and their nearly unarmed condition at independence. Because of the paucity of weapons in the system as a whole, the additions that were made in these years radically alter the average correlation. Beginning in 1959, the curve shoots rapidly upward, going from below zero to

TABLE 3.2
Total System Correlation Matrix: Naval Weapons

Category	Guard boats	Patrol boats	Corvettes	Amphibious vessels
Guard boats	1.0000			
Patrol boats	0.4104	1.0000		
Corvettes	0.4509	0.3107	1.0000	
Amphibious vessels	0.5243	0.6355	0.3454	1.0000

NOTE: 11 of the 46 states in the sample are landlocked and do not have navies.

nearly 0.5 in 1966. Curiously, it then dips, ultimately losing over a quarter of the previous years' gains. Not until 1973 do we see the beginning of a slow recovery. The initial surge in average correlation reflects the military reorganization that usually followed on independence. As we saw in Chapter 2, the military forces in most states were ill-prepared for independence. Their initial average scores were therefore exceedingly low (actually negative in 1959), and the subsequent rapid rise not unexpected, with its strength also reflecting the fact that the increasing stocks of weapons in the system (Figs. 3.1-3.4) made it easier for all categories to be filled and some level of balance obtained.

The reasons for the subsequent dip and recovery are not so readily explained. The dip cannot be ascribed to unbalanced imports of arms at the system level; as noted earlier, these were remarkably linear. Two factors seem to explain why the system average correlation failed to climb beyond the 1966 values. One is that the smallest states in the system began their arming process later and carried it out more slowly than the others. Because their size did not permit arming across the board all at once, they also acquired weapons sequentially in different categories. This means that when they started to arm themselves in the mid-1960's, their actions drove down the overall average balance.

A more important explanation for the "U" shape of the correlation plot after 1966 can be seen in Figures 3.7-3.9, which show the growth in total-system naval, land, and air weapons by category across the 25

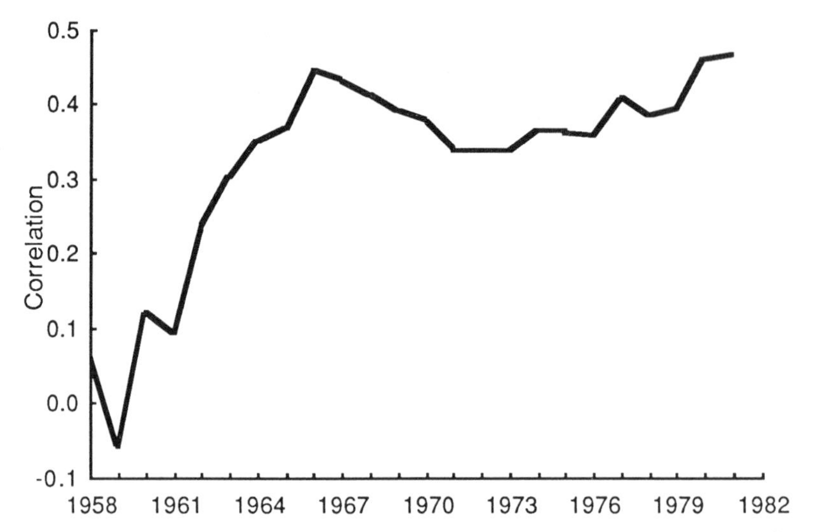

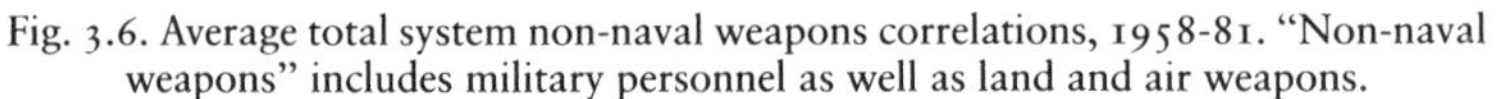
Fig. 3.6. Average total system non-naval weapons correlations, 1958-81. "Non-naval weapons" includes military personnel as well as land and air weapons.

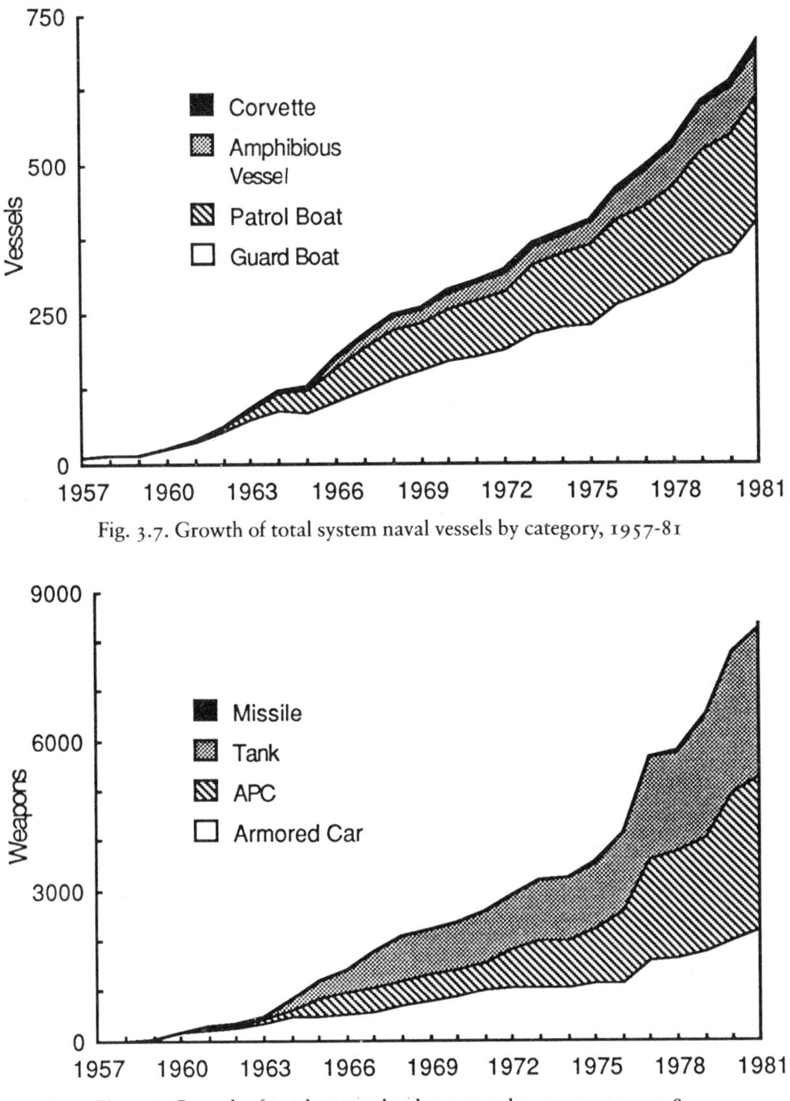

Fig. 3.7. Growth of total system naval vessels by category, 1957-81

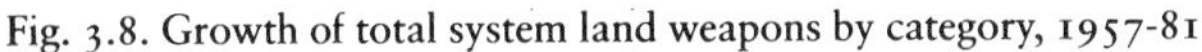

Fig. 3.8. Growth of total system land weapons by category, 1957-81

years of the study. As they clearly demonstrate, within the linear total-growth pattern, the composition of all three arsenals changed substantially over time. By 1965, many of the new states had begun their first re-equipment cycle. The colonial officers had largely left by then, and for the first time, the new states could break away from colonial pat-

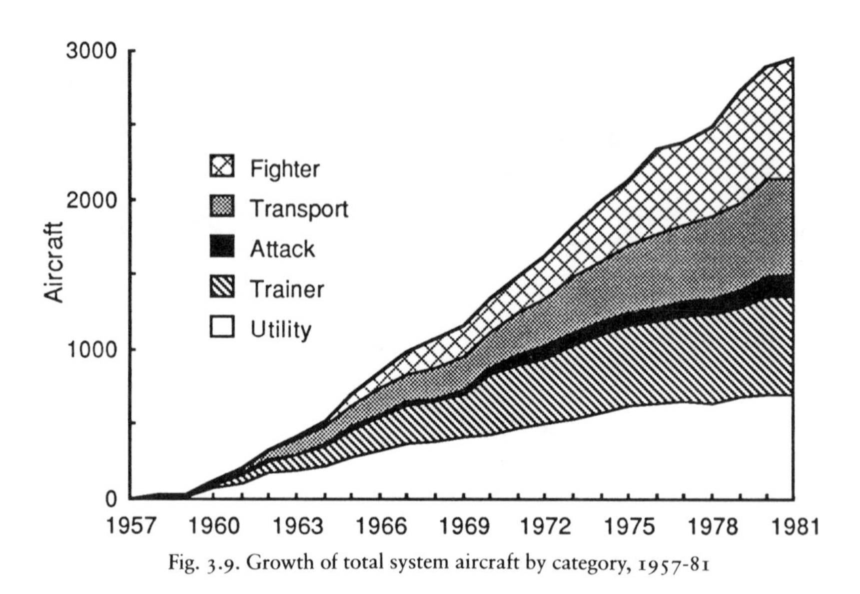

Fig. 3.9. Growth of total system aircraft by category, 1957-81

terns and structure their forces to their own choosing. By the middle 1960's we see significant numbers of high-capability weapons, and the total dominance of low-capability systems evident during the earliest post-independence years begins to wane. Thus we see the total number of high-capability patrol boats growing from essentially zero before 1961 to 30 percent of all vessels by 1981; tanks and APCs, originally greatly outnumbered by low capability armored cars, dominating the totals by the end of the period; and nearly all the growth in aircraft after 1970 coming in high-capability fighters and transport planes, which were hardly to be found earlier in the period.

The states with middle-sized forces tended to reduce the balance of their forces during re-equipment cycles because they could not afford to replace or add weapons across all categories at the same time. The states with the largest forces (and the greatest impact on the system totals), such as Malaysia, Kuwait, Somalia, and Nigeria, did not become markedly less balanced during the mid-1960's, so the average values do not entirely collapse. Once the initial stages of re-equipment were completed, the weapons stock totals were sufficiently high in each category that the correlations were no longer very strongly affected, so the average begins to climb again, though still held down by the effect of the very small states mentioned above.

Weapons intercorrelations can also be used to examine the acquisition patterns of individual states. Figure 3.10 is a histogram showing the average of the military personnel and non-naval weapons correlations by state for the whole time span of the study. These mean correlations are one indicator of how balanced a state's military force is, averaged over the entire period of its independence. The histogram is quite spread out, and the variation in average correlation is considerable. The explanation for some of the distribution is quite straightforward. All of the ten states in the lowest scoring range are very small states, which acquire weapons in smaller numbers and at a slower rate than others, and are thus likely to upset the balance of their forces more readily with each new acquisition. This is especially true for the seven island states, which maintain very low stocks of armored fighting vehicles.

Above the lowest group, however, no clear pattern emerges. South Yemen, which thanks to Soviet aid has a considerable military force, shares the next range with Guyana, which has next to nothing. Likewise Bangladesh, one of the largest countries in the study, shares the third range up with Togo, one of the smallest. Malaysia, Nigeria, and Kuwait, all of which have among the largest military forces in the group, rank in the middle; each shows an average correlation between

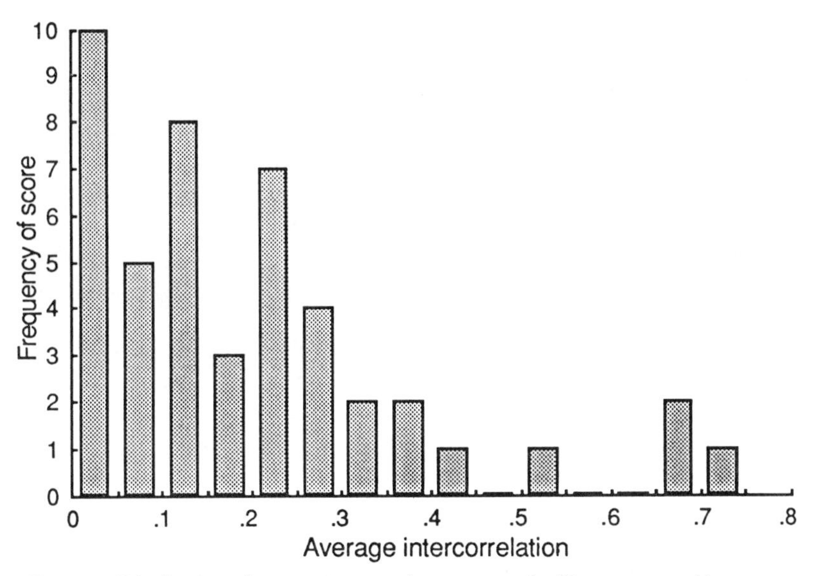

Fig. 3.10. Distribution of average non-naval weapons and military personnel intercorrelations for the 46 states, 1958-81

0.29 and 0.35, which is above smaller Mali but below Guinea. Singapore shows the most balance among categories, clustered at the top with Algeria and Zambia. While size and sophistication of arsenals might explain the presence of Algeria and Singapore in this range, it is not clear why Zambia should do so much better than its neighbors. Clearly, the balance of weapons varies among these states, at least by the rough measure of it used so far, and with the exception of the very smallest, it does not seem to correlate highly with size. This adds confidence to the decision to include balance as an element in calculating military capability, since it seems to tap something different than merely aggregate force size.

Order and Diversity

A first cut at the military capability data has revealed enormous diversity among individual states within a system of striking order and regularity. The system displays uniform growth, and indeed, although the individual states did not grow continuously, they almost never reduced their forces from one year to the next. Even when domestic economic conditions were very bad, as they were for Ghana and several other West African states when commodity prices collapsed in the mid-1960's, the individual military forces remained at least of constant size and never shrank.

The evolving structure of the new state forces can be traced as the balance among categories changed with time. It had been explicitly the plan of the French, but implicitly the plan of the other colonial powers, that these new states would remain armed only for internal security purposes and under the military protection and tutelage of the former metropole. This in fact was largely the case until the late 1960's, when the system totals graphically show the rising share of the high-capability weapons.

In the first decade of the period, about two-thirds of the 46 states embarked on weapons acquisition programs that rapidly increased the number of more advanced weapons in the system. From this we can infer that inequality in military capability was increasing among system members even though the growth of the system as a whole was unperturbed, but at what rate and to what extent remains to be determined. Once these questions have been answered, the variation among states in capability acquisition can be made explicit, and the search for the determinants of that variation begun in earnest; for the system-level regularity cannot be fully understood until the behavior of its individual members is accounted for.

FOUR

Capability and Development

Many methods for measuring military capability have been employed in the past, including general indexes of economic power and industrialization, predictions based on previous performance in combat, and sums of hardware and military units. But as noted, none of these are suitable for a set of states so large and varied as the sample under study. What is wanted is an index that can transform annual national data on military personnel and 13 categories of weapons stocks into a single annual military capability score in such a way that it takes into account both the quantitative and the qualitative components of capability discussed earlier. That is, scores should rise the greater the stockpile grows and the more balanced it is. These demands are fairly strict; they eliminate simply summing across all categories and require that the method chosen accommodate data in a variety of units (people, tanks, etc.) spanning widely different ranges of values (1,000 to 763,000 for military personnel, 0 to 900 for tanks, etc.).

Index Construction

Because the approach taken here to constructing a military capability index has not been used before, it is important to describe the construction process in some detail. The algorithm most closely matched to the requirements described above is Conjoint Measurement III (CM-III; Lingoes 1973). In employing CM-III the categories need have no known linear relation to each other or to the underlying capability dimension one wishes to measure, but the data in each of the categories to be scaled must be at least rank-ordered. The categorized military capability data set meets this requirement.

CM-III transforms the different variables so that they are linearly related both to each other and to the capability dimension, and then calculates a single "best score" that measures rank on that underlying dimension. It does this by differentially shrinking and stretching the intervals between values until an optimal scaling is achieved (i.e., when linearity between categories is maximized), subject to the constraint that if a variable fed into the algorithm has a higher score for case A than for case B, then the final transformation of the variable will also have a higher value for case A than for case B. The transformed scores will thus be monotonic with the original data; that is, the rank-ordering of cases on each dimension will remain unchanged, and the combined capability score will be linearly related to each. A detailed description of the operation of the CM-III algorithm is given in Appendix B.

The only significant problem with using CM-III is that the data set includes several landlocked states without navies. If the algorithm were run on the entire data set at once, these states would be penalized in their capability score. In order to overcome this problem, the index must be constructed in three stages. First, the algorithm is run for the entire set of states on only non-naval weapons and a set of composite capability scores obtained. The process is repeated for naval weapons for just those 621 country-years that did not involve landlocked states. It is then possible to calculate the relationship between naval and non-naval scores for the coastal states, and from this relationship derive an artificial navy score for the landlocked states based on what they would have had if they had behaved like coastal states with similar size forces. Adding the artificial navy scores provides a complete list of 859 scores in both naval and non-naval categories, which can then be run through the algorithm again to produce a final composite total.

The algorithm calculates a new score for each variable plus a composite. In this case, CM-III required two iterations to accomplish the task for the non-naval categories. The mean correlation between the original ten non-naval variables was 0.58 for all 859 cases; it fell to 0.44 for the ten transformed variables plus the eleventh composite one. For the naval scores the program also went through two iterations and the mean correlation went from 0.45 to 0.40 for the 621 coastal cases. Note that the average correlation actually declined a bit in each case. This is an indication of the degree to which the original correlations depended on outliers for their values. As the algorithm proceeded, it stretched and shrunk intervals and in effect reduced the impact of outliers. Once this reduction began to subtract more from the mean correlation than the other actions of the algorithm added, the CM-III program ceased iterating.

That the average correlation among the transformed scores remains low is no surprise; it reflects the low correlations of the actual totals in the real world. The newly computed composite score does better, however, and this is what is more significant. The average correlation between the composite non-naval index and the transformed scores is above 0.7 and for the naval category it is above 0.8. These are respectable values and indicate that the composite indexes are an improvement over any one individual category or average of category scores.

CM-III was designed for use even in cases where it is not clear whether a variable should add or subtract from the total score as its value increased. Therefore, if any negative correlations are discovered, the algorithm reflects the values so that less will be better for that variable. In other words, if it turned out that tanks usually went down when every other category went up, the algorithm would assume the more tanks, the lower the capability score. As was expected, the algorithm did not reflect any of the variables in either the naval or the non-naval case; more was always better, which was one of the original assumptions of the index-construction process and was expected from the patterns in the raw data.

Combining the naval and non-naval scores into a single index is quite straightforward. Using least-squares regression, the relationship between naval and non-naval composite scores for coastal states was found to be (naval score) = 0.657 (non-naval score) − 0.055.

With R^2 equal to 0.424, the fit of this formula to the data is certainly not perfect (1.0 would be a perfect fit; zero would mean no relationship), and it can be expected to reduce the accuracy of the final index by introducing errors into the estimate of the scores for landlocked states. However, least-squares regression produces the maximum likelihood estimator of the true relationship between the two variables and is the best available statistical tool for the job. It is preferable by far to the alternative of ignoring landlocked status and simply giving those states a zero score in naval categories.

When the CM-III program was run a third time, the non-naval scores were used for ten categories and the naval scores for four, to preserve the original relative weight of each component. Of course, the initial matrix for this run showed a very high average correlation ($r = .840$), and the program proceeded for 14 iterations before halting when r exceeded 0.900. The average correlation of the final index with the last input scores was over 0.968. The algorithm operated as advertised, and the result was a list of 859 military capability scores covering every state and year of the study.

A few quick checks confirm the face validity of the index-construction process. Unchanged manpower and weapons figures over time should result in tie scores, and in fact they do. In one case, tie scores also permit an examination of the impact of using artificial navy scores for landlocked states. Mauritius had no military forces for the first 11 years of its independence. It has the same military capability score for all those years and it is, appropriately, the lowest score of any in the data set. Lesotho also had no military forces for the first few years after independence, but Lesotho is landlocked, and so its naval scores were artificially generated. Since the formula for doing this does not produce a perfect fit, Lesotho received a slightly non-zero navy score even though its non-naval score was zero. The results of this process are no real cause for concern, however, because even with its artificial navy score, Lesotho ranks next to Mauritius and below the lowest real non-zero scoring state for those years.

Since most states gained independence with very few weapons and military personnel, one would expect to find more scores at the lower end of the scale than the upper. Also, since it is not possible to have less than zero in any category, there should be more room for variation at

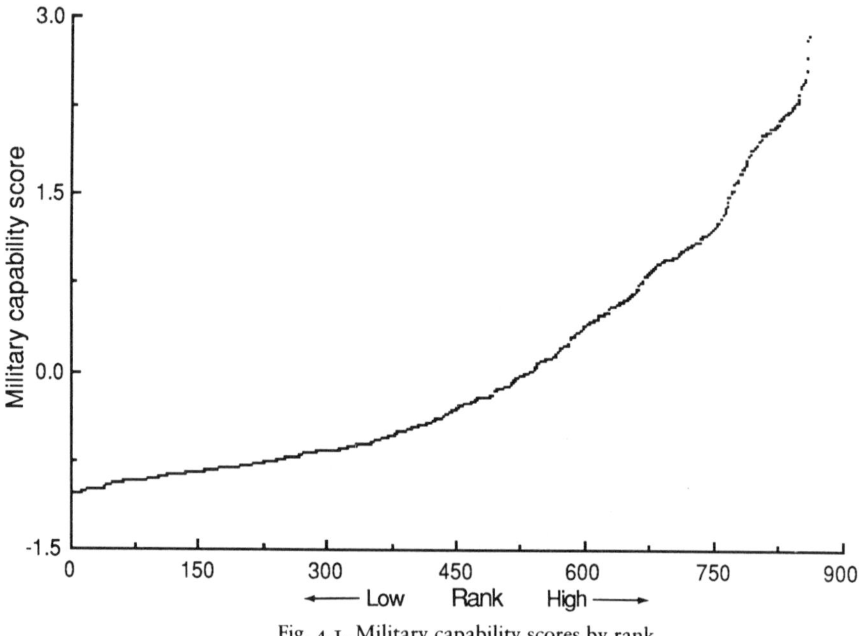

Fig. 4.1. Military capability scores by rank

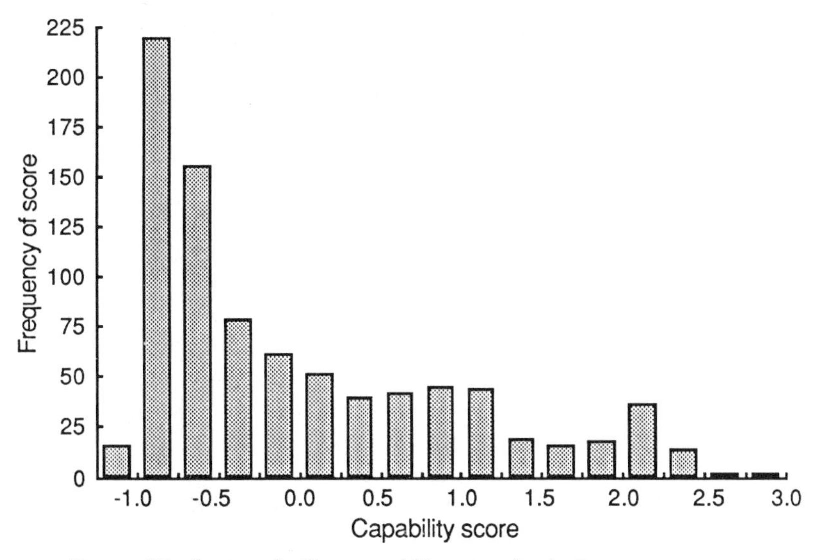

Fig. 4.2. Distribution of military capability scores for the 859 country-years

the upper end of the scale than at the bottom. The actual distribution of capability scores fits these expectations.

Figure 4.1 plots the capability scores on a uniform scale from least to greatest. The scores range from about −1.5 to 3.0 (the minus signs are of no importance; they are a result of the way the algorithm operates). The figure shows that the scores tend to spread out as they get higher. Note that the relationship is not linear; the slope increases with rank as the room for variation increases.

Figure 4.2 is a histogram that displays the third dimension of Figure 4.1, the distribution of scores from lowest to highest. While the small number of scores in the lowest category reflect essentially zero military forces, in nearly all country-years there are found at least some military weapons and personnel. Note, however, that the scores are heavily concentrated at the lower end of the scale. In fact, nearly half of the scores are in the bottom 20 percent of the scoring range, and only a tiny fraction of the total are found near the top. Clearly, there is considerable variation in the progress among these states in military capability growth.

Figure 4.2 provides one perspective on the growing inequality in military capability among this group. All started out weak, and while some stayed weak—so weak that the low scores dominate the total—others made considerable progress. Figure 4.3 provides a precise measure of the course of this process over the period. It is a plot of the standard deviation in score on the composite military capability index during each year of the study. Standard deviation measures how dispersed the scores are around their mean: a low standard deviation indicates the scores are concentrated; a high one indicates they are widely dispersed. At the start, the standard deviation jumps around as the system grows from two to twenty-one members in just three years, but then it settles into a monotonically increasing trend. The correlation between year and standard deviation is in fact 0.94, and higher than that if measured from 1960, when the system size is more stable. The slope decreases with time, nearly leveling off after 1977, showing that the growth in inequality is decreasing. Put another way, to an increasing extent all new states are participating in an expansion of military capability, so that averaged over all states the degree of inequality has been leveling off, but some have gone much further in expanding their capabilities than others.

Figure 4.4 is a scatter plot of all military capability scores by year, and it shows most clearly the pattern that developed. After 1960, the states diverge on military capability like the spokes on a wheel, and though some remain almost entirely unarmed, the annual scores are spread out quite uniformly across the chart by the late 1970's. One can

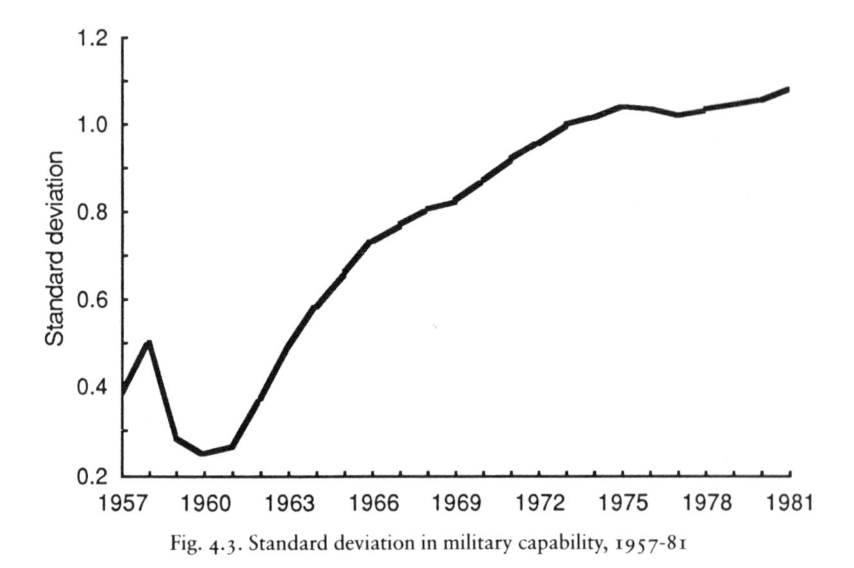

Fig. 4.3. Standard deviation in military capability, 1957-81

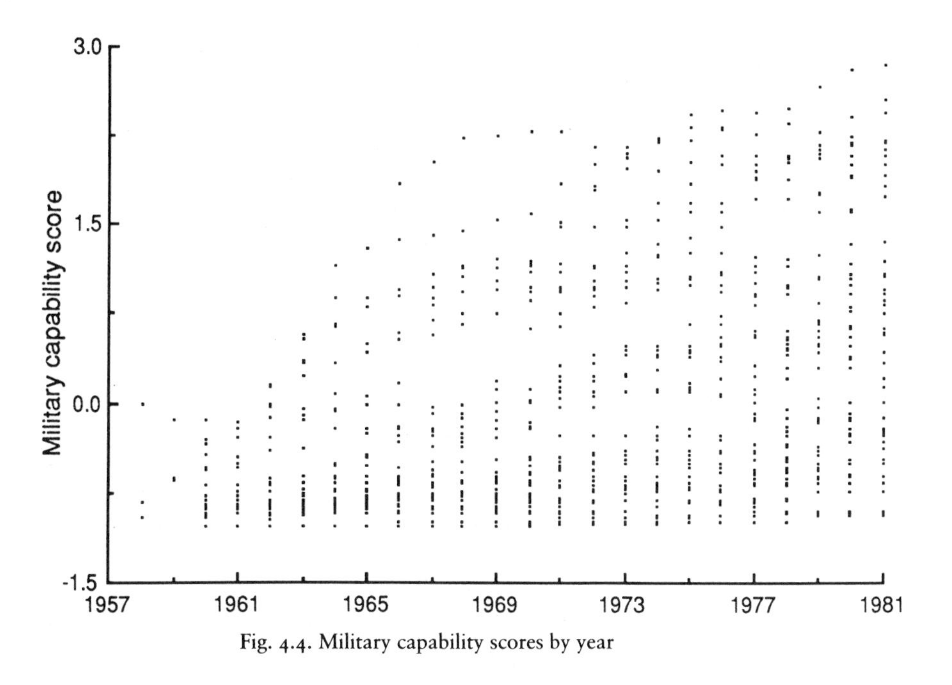

Fig. 4.4. Military capability scores by year

also discern the curvature of the standard deviation plot we saw in Figure 4.3 as the rapid rise in the highest scores begins to slacken after about 1970. These scores provide the measure needed to examine the role military capability has played in the development of these new states.

Measuring Development

Like military capability, development is a complex concept, defined on numerous dimensions and measured differently on each one. Some very interesting methods for operationalizing and measuring it have been proposed over the past two decades (e.g. Pye 1965; Huntington 1971; Rokkan & Eisenstadt 1973), but none is useful in every application.

Development can be viewed as an economic process that operates to increase aggregate or per capita wealth. It can also be viewed as a political process in which an increasing fraction of the total population identifies with or becomes subject to control by the central state. A concept of development most relevant to this research would combine both views and define it as a process in which governments (or their ruling elites) acquire the capacity to create and deploy national power for their own purposes. Defined in this way, development reflects national capacity, and combines the economic notion of aggregate wealth with the political notion of control over that wealth. It provides an index of potential power, that is, an upper bound on the capacity of a state to mobilize resources for whatever purpose, military or civil, when the need arises. Development as national capacity meshes well with the feedback model of European development, but the problem is that there is no particularly satisfactory way to measure it.

Approaches to development as national capacity can be grouped into three general categories. First, there are indexes that employ single aggregate economic indicators in an attempt to measure the sum total of national wealth. The most widely used of these indicators is gross national product (GNP), defined as the value of all goods and services produced in a nation during a 12-month period. GNP provides the most comprehensive measure of the capacity to pay for armed forces (or anything else) and an indirect measure of the capacity to aggregate national production efforts into a pool of usable capability. However, GNP does not provide any measure of the effectiveness with which governments are able to mobilize capacity and direct it toward a common goal or a direct measure of any other political development criteria.

Some approaches to indexing development have made an effort to account not only for the wealth produced in a state, but also for the government's ability to channel that wealth to its own purposes. For example, Organski and Kugler (1980) created an index of efficiency that, when multiplied by GNP, measures the resources a government is capable of directing at any desired application. Organski and Kugler argue persuasively that one cannot assume political development will always keep pace with economic growth just because it did so in Europe, and that since one cannot automatically infer political performance from economic performance, measures of national wealth may not, by themselves, be adequate measures of national power potential. To account for political development, they weight GNP with an efficiency index created by dividing actual tax revenues by tax capacity. Tax capacity is defined as the revenue predicted to be collected given available resources relative to other nations in the system, and it is calculated by regressing the actual tax ratio against economic indicators of differences in resource base across all possible country-years in the time span of interest. The quotient of real tax revenues over tax capacity multiplied by GNP thus provides a measure of the potential resources a government has at its disposal for whatever purposes, military or otherwise, it chooses.

A third approach to measuring development has involved attempts at multidimensional scaling that generally include indicators of population size, political integration, and industrial production, as well as an aggregate measure of national wealth. An example is the capability index drawn up by Singer, Bremer, and Stucky (in Russett 1972), which ranks each nation by its iron and steel production, energy consumption, total population, urbanized population, and military expenditures and personnel to produce a final relative scale. It is very difficult to employ cross-temporally, however (the scaling is done by year, and the sample must remain constant), and is of questionable value for states that do not produce most of their own weapons. Indexes of this type can be useful for comparing nations of similar size and level of development, but they present considerable difficulties when used with a sample as broad as the 46 states under study.

In any case, since such indexes concentrate on the heavy-industry sector, this approach is quite unsuitable for our purposes. None of the 46 states is even a minor arms producer, so their levels of industrialization have no direct relationship to their weapons acquisition programs. Although industrial output obviously affects national wealth, for states that import all their arms it is the level of wealth more than its source that matters. In addition, because the sample ranges from two to 46 states over the 25 years, the Singer-Bremer-Stucky technique of using a new scale each year is simply unworkable.

The Organski-Kugler index is ideal in theory, but less than ideal in practice for this set of states. For one thing, data on taxation are unavailable or inaccurate in a number of cases. For another, in many of these states, most of the national income derives from the sale of natural resources. For the governments of these states, income is extracted not through direct taxation, but primarily through royalties or export taxes on oil, copper, bauxite, and the like. Although Organski and Kugler propose ways of handling this problem, in the end these methods proved to be difficult and produced inconsistent results.

Therefore, though I shall refer to various indicators of development, including the Organski-Kugler efficiency index, in the pages that follow, the core statistical analyses will be based on GNP. Its weaknesses notwithstanding, GNP does provide an analytical baseline onto which other concepts and indicators can be grafted as necessary. Given the large number of countries and the long time span of the study, such a universally applicable baseline is a necessary part of the general framework, whatever is added later.

Global Patterns

Data can be grouped for analysis in different ways. We have already looked at simple system totals for military stockpiles and GNP, and found both a remarkable degree of linearity in each and a remarkable degree of correlation between them (Fig. 3.5). Other levels of aggregation permit conclusions about different facets of the development-capability relationship, but care must always be taken not to infer a relationship at one level of analysis from the results of an investigation conducted at another. With the military capability index complete, we can now examine the data with greater clarity at all levels of analysis.

Figure 4.5 plots the growth in average military capability and average GNP from 1960 onward. The growth in average capability is even more linear than the system growth in weapons stocks, correlating with time at an amazing 0.998 (for GNP the figure is 0.972). The essentially perfect correlation with time is not an artifact of the indexing technique; CM-III does not "know" what country or year it is calculating a score for and does not attempt to make the index linear across either dimension. The observed linearity is quite unexpected; the con-

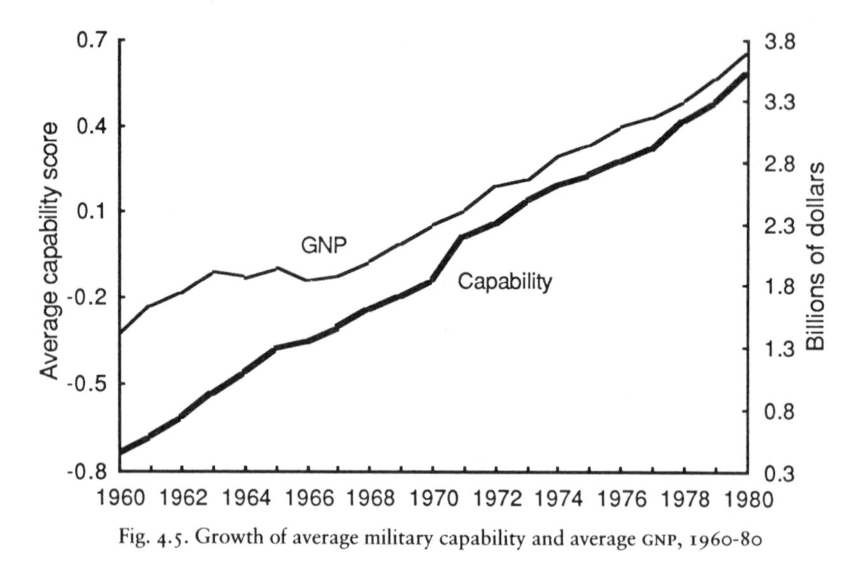

Fig. 4.5. Growth of average military capability and average GNP, 1960-80

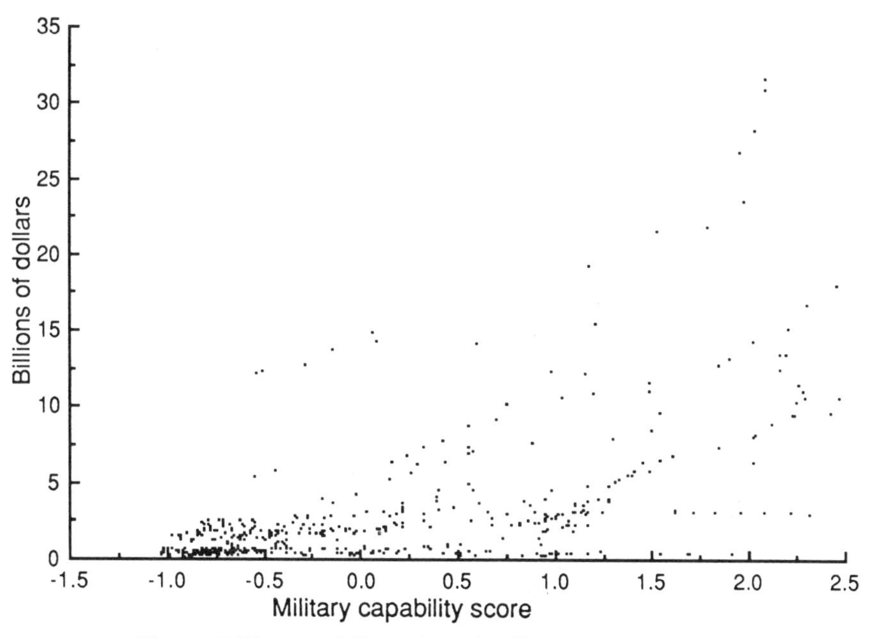

Fig. 4.6. Military capability and GNP for all country-years

sensus in the literature has always favored accelerating growth in military capability. This new figure can tell us more than Figure 3.5, because using averages compensates for the changing number of states over time and prevents a few large states from dominating the totals. We can say with certainty, therefore, that the average military capability of new states has been growing linearly and (especially from 1966 onward) in parallel with average GNP. We cannot, however, say anything yet about individual states. To do that we must look at the country-year scores themselves.

Figure 4.6 is a scatter plot of GNP versus military capability for all country-years in which data are available. The correlation of GNP with military capability is 0.625, indicating that some relationship between GNP and military capability is found among system members as well as for the system as a unit.

Nearly all the points lie in the lower left half of the figure. The principal exception is the small cluster of points at the top—these turn out to be the scores for Kuwait. Kuwait is a major oil producer and has one of the highest average GNPs in the group. But Kuwait is so wealthy for its small size ($14,800 per capita in 1975, among the highest in the world) that it cannot possibly achieve the military capability one would predict for it on the basis of the rest of the group. For instance, with over 24 military personnel per 1,000 population, Kuwait has the highest military manpower ratio in the sample (more than twice that of South Yemen, the next highest, and also 2.5 times that of the United States), but this still leaves it with a force of only 25,000 men. Kuwait does not seem to be a country behaving any differently from the rest, just one that has run up against the physical limitations of its size.

Removing Kuwait still leaves a few points in the upper right-hand corner with scores so high on GNP that they compress the rest of the points into a relatively small area and make patterns difficult to discern. These points belong to Nigeria and Algeria, which join Kuwait at the top of the average GNP list. Like Kuwait, they have so much oil money that they have difficulty acquiring military capability fast enough to keep up with their wealth, especially after the oil price rises of the early 1970's.

By removing the points for all three countries, we can spread the rest of the country-year points out to see whether a pattern can be detected. This has been done in Figure 4.7. Although the correlation co-

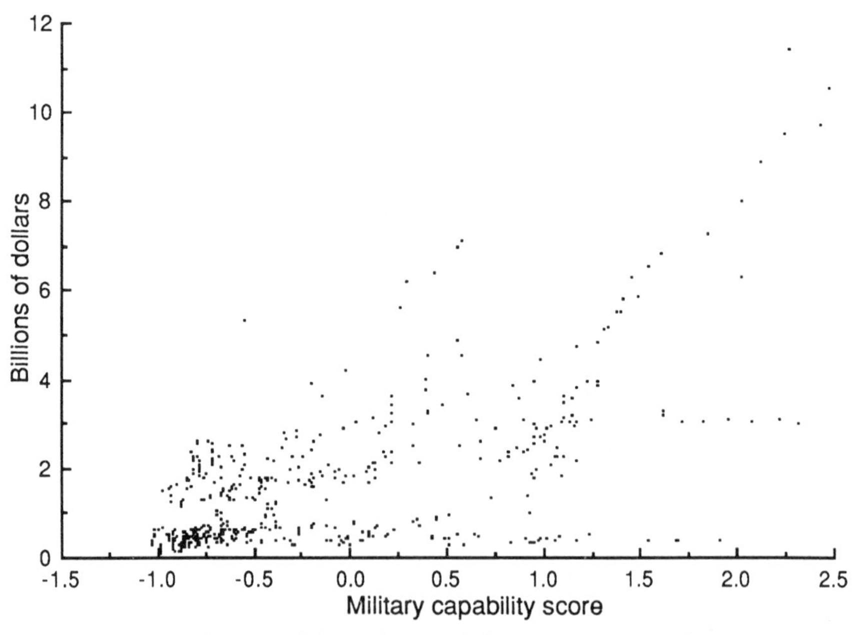

Fig. 4.7. Military capability and GNP excluding the country-years for the three wealthiest states

efficient goes up slightly with this step, to .664, the three states were not removed because their behavior was different from the rest, and the important point is the pattern that emerges. Figure 4.7 shows clearly a weak monotonic relationship between military capability and GNP. That is, the upper left-hand quadrant of the plot is empty, showing there are no states that score highly on wealth or development but poorly on military capability. The bulk of the country-year points fall roughly on the diagonal, but a significant fraction lie below it. These points represent country-years in which despite little in the way of development, substantial levels of military capability were amassed; wealth turns out to be a sufficient but not necessary predictor of military capability.

One would expect that foreign military assistance would be a major determinant of military capability, and indeed that it might well explain, in great part at least, the off-diagonal points; but as noted earlier, reliable data on the costs of hardware transfers are not generally available. It turns out, however, that some very simple coding decisions provided the key to understanding the poor but powerful country-years.

Recall that in Chapter 2 we saw that those states that accepted the Soviet Union or (to a lesser extent) the PRC as a patron at some point after independence tended to receive higher levels of military equipment than those that did not. By coding each country-year for the occurrence or non-occurrence of a patron shift away from the former metropole (the shift defined as beginning the first year in which a majority of imported weapons came from the USSR, the PRC, or an ally of either), one can determine whether the military aid from those countries explains the off-diagonal cases. And in fact when these patron-shift cases are excluded, all but a dozen or so of the off-diagonal points disappear.

Essentially all the remaining cases turn out to be the beneficiaries of French military assistance in the 1970's. After 1971, France reinvigorated its military ties with several of its former colonies, and increased military supplies followed. The flow of military aid accelerated after the mid-1970's as French concern about the growing Soviet and Cuban presence in Africa mounted. For the first time, sophisticated jet aircraft were transferred to small African states like Gabon, apparently without a requirement for cash payment. When the increased French aid after 1971 is taken into account and those country-years affected are removed from the group, the scatter plot displayed in Figure 4.8 results. The correlation between military capability and GNP is a most remarkably high 0.854.

The degree of strong monotonicity evident in this last scatter plot is

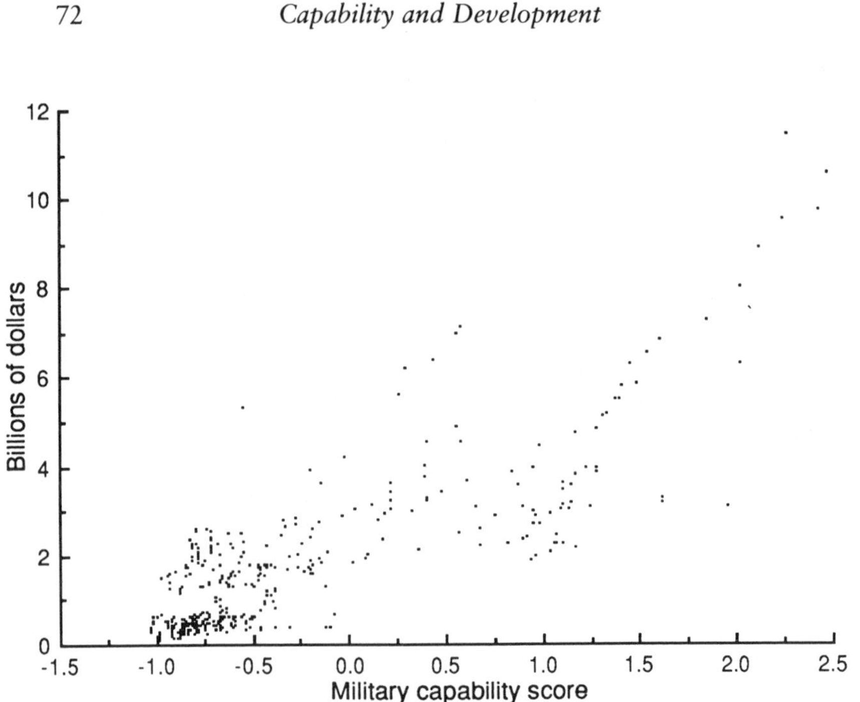

Fig. 4.8. Military capability and GNP excluding the country-years for the three wealthiest states and the beneficiaries of military patrons

surprising in light of what has been said in earlier chapters about the wide variety of defense circumstances among these 46 states. Although individual nations might display a strong capability-GNP relationship, that in no way suggests that such a relationship should pertain to the country-years as a group. That is, there is no reason to assume that a state in one region would maintain the same level of military capability as a state of similar means in another region with quite different defense requirements. On the contrary, one would expect that states in peaceful regions would arm themselves relatively less than those in hostile regions, quite apart from their respective levels of wealth.

Nevertheless, GNP and military capability are strongly related at the country-year level, and when the effects of great wealth and heavy patron support are taken into account, the pattern is quite clear. Were we able to be more precise in accounting for foreign aid, it is quite likely that we would find an even stronger relationship between development and military capability. For instance, many of the points in Figure 4.8 at the lower right belong either to Zaire, which got far more U.S. aid than most African states, or to Zambia, which diversified its military suppliers to include the Soviet Union after 1974, as the civil war in Rhodesia grew in scale.

Interestingly, categorizing the country-year data in other ways has very little impact on the capability-GNP relationship. It does not, for instance, vary with the identity of the former metropole, although with the factor of French aid to its former colonies in the second half of the period, coding by metropole alone does raise the correlation coefficient slightly. Nor does the relationship vary between revolutionary states and those that achieved independence peacefully, although here again, since Algeria was both a revolutionary state and one of the three wealthiest and all the revolutionary states receive Soviet arms, simply excluding the revolutionary group raises the coefficient somewhat all by itself.

Calculations of an efficiency index done by Organski and his colleagues are available for about a third of the study country-years. GNP and GNP weighted by this index are correlated about equally with military capability. GNP alone does slightly better as the patron-shifters are excluded because many of the remaining states derive most of their income from royalties on natural resources, a factor the weighted index was expected to have difficulty with. In general, however, their similar behavior adds confidence to the decision to depend largely on GNP as a measure of development.

Rates of Change

We might reasonably suppose that the nature of the relationship between development and military capability would show up most clearly when comparing their rates of change, because the great variation in defense circumstances among the sample states ought to hide any relationship between absolute values. In fact, there turned out to be a very strong relationship between the absolute values, and as we shall see, a relationship between the rates of change is hardly evident at all.

The first clue to what is happening with rates of change can be seen in Figure 4.9, which plots the average of all the individual growth rates in military capability and GNP for the 20 years between 1960 and 1980. The average annual rates are much less stable than the annual system averages, which grew linearly over this time period (Fig. 4.5); and though the two growth rates move very loosely in unison, the tight relationship shown by the absolute values is missing.

At the country-year level, the relationship between rate of growth in

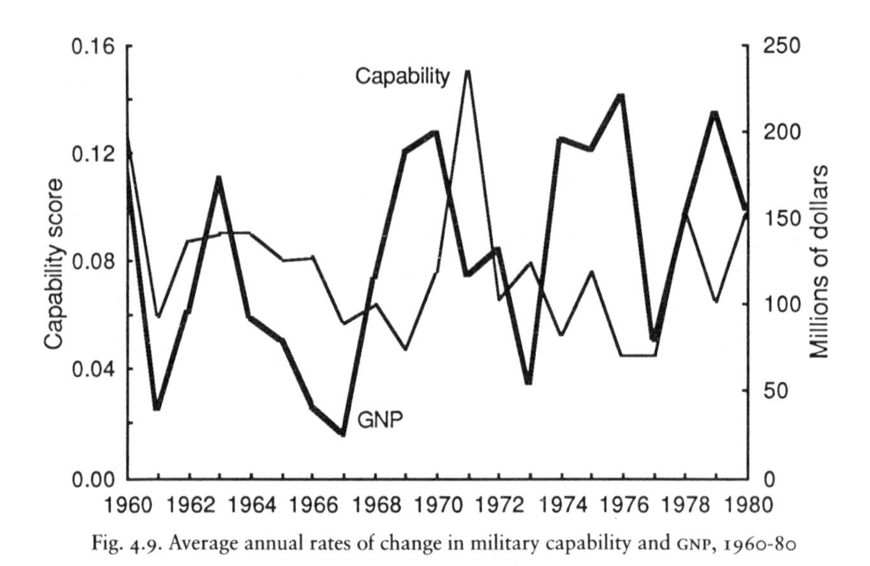

Fig. 4.9. Average annual rates of change in military capability and GNP, 1960-80

military capability and GNP disappears entirely; the correlation coefficient between them is 0.070. An essentially zero correlation between rates of change after such high correlations were obtained between absolute values is quite startling, but as before, Kuwait, Algeria, and Nigeria are such outliers on the GNP index that their presence makes any pattern in the cluster of points difficult to discern.

Figure 4.10 displays all country-year rates of change with the exception of Kuwait, Algeria, and Nigeria. Removing these three countries has had little effect, and the plot is little more than a random blob with a correlation coefficient of only 0.132. Clearly, the problem is not that outliers distort the general correlation picture. Neither, in fact, does patron support. The patron-shift concept that was so useful in understanding the absolute values makes no difference for the rates of change, raising the coefficient to only 0.177.

At least one, if not both, of the variables must behave quite erratically, then, when measured at an annual rate. This is confirmed by Figure 4.11, which shows the standard deviations (divided by their respective means to control for different units) plotted over time for military capability, GNP, and the annual rates of change in each. The value for military capability nearly quadruples over the period (as shown earlier in Fig. 4.3), and the value for GNP nearly doubles, but their behavior is almost lost in the chaos of the values for rates of change. The

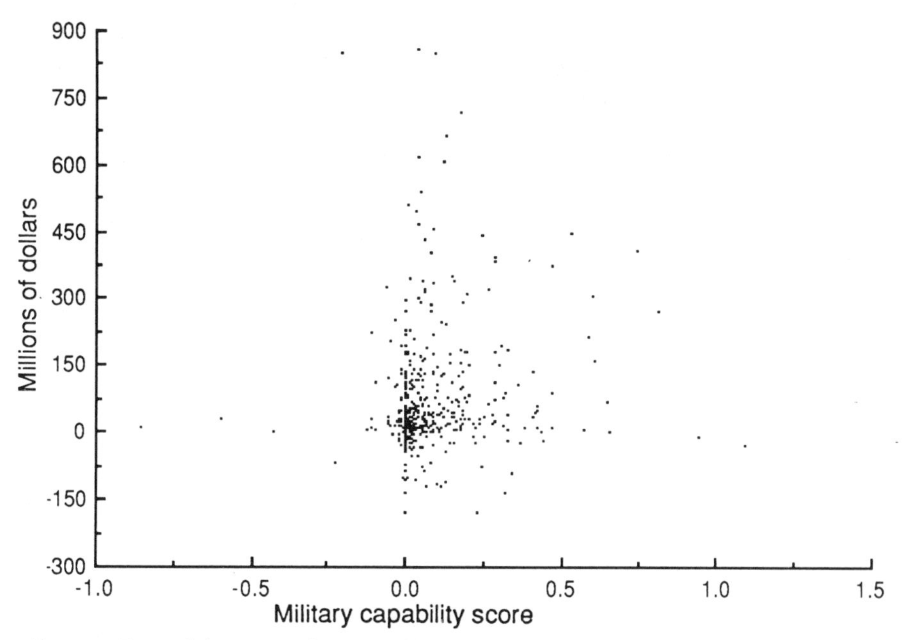

Fig. 4.10. Rates of change in military capability and GNP excluding the country-years for the three wealthiest states

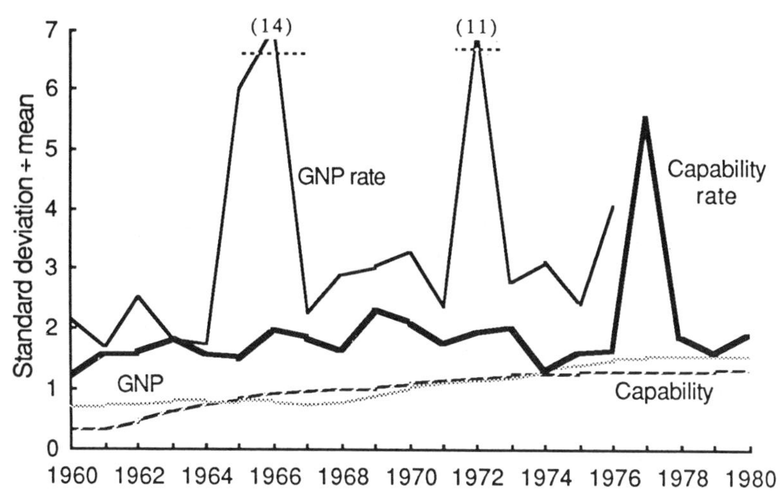

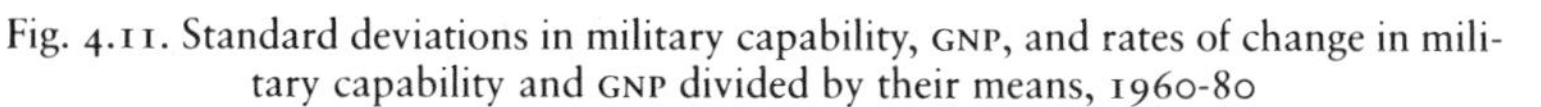
Fig. 4.11. Standard deviations in military capability, GNP, and rates of change in military capability and GNP divided by their means, 1960-80

standard deviation in rate of change in military capability averages about twice that of the static indicator, but is consistently less than the deviation in the GNP rate, which is by far the most erratic. In two cases, it is actually off the scale of the figure, exceeding 14 in 1967, owing to a sharp rise in commodity prices, and exceeding 11 in 1973, when the first large oil price increases took place. These figures show that the scores on rate of change not only are more spread out than the static indicators, but fluctuate much more radically in their distribution from year to year as well.

If the erratic annual behavior of capability growth and, especially, GNP hides any relationship between them, it may be that over longer periods their progress is more stable and perhaps also more synchronized. In this case, moving averages could illustrate a relationship hidden in the annual figures. Recalculating the rates of change in each indicator as five-year moving averages results in a mild increase in country-year correlations across the board. GNP and capability correlate at 0.247 for all available country-years and slightly higher (0.347) with the three wealthiest states removed. But these figures are nowhere near the absolute values. A way to confirm that the correlation increase is due to an increase in the indicators' stability is to calculate the average standard deviation in each annual rate of change and moving rate of change by country and then divide it by its respective mean (to control for units). The results are as expected: the five-year averages are more stable than the one-year statistics, though only slightly, and as just noted, using them produces only a mild increase in correlation coefficient.

Interestingly, when the two halves of the time period are calculated separately, the country-year correlations between GNP and military capability are slightly higher for the second half. The rates-of-change correlation remains essentially zero in both cases. It may be that whatever distortions the colonial period created disappear gradually with time and so an underlying relationship emerges as the post-independence period lengthens. Returning to Figure 4.11, we see that the states were much more unequal in GNP in 1960 than in capability (all were weak; not all were poor). By 1980, there is more inequality in capability than wealth, but the gap is narrower and probably greatly affected by patron support.

No such time-dependent patterns emerge among rates of change; it turns out they have limited utility as indicators of any relationship between military capability and development. If a very long period is available for analysis, then moving averages calculated over a long span can prove fruitful. At this point, however, a ten-year moving average calculated for the 1960-80 period would provide only ten annual data points, an excessively narrow band for checking on the consistency of any of the relationships over time. What the erratic behavior of the rate-of-change data does suggest, however, is the need for caution in inferring any causal ties from the very significant static-data results.

This chapter has assaulted the reader with a swarm of numbers and charts whose connection to our original focus may not always have been perfectly clear. Accordingly, this a good place to stop and review what the data have shown and how the findings relate to the fundamental questions of this research.

Recall that we want to examine the relationship between military growth and development. Specifically, does a feedback relationship obtain in which the search for military power is a driving force in the development process? To this point, the data analysis suggests that the answer is no in the present case. Both interesting and surprising is the close relationship between GNP and capability at the system and country-year levels of aggregation. Nothing about the country-year scatter plots suggest that any external factors except patron support drive the growth of military capability. Since military capability is supposed to be created in response to the external environment, the closeness with which it is correlated to GNP at the country-year level without regard to external security is quite unexpected; and the regularity with which system-level military capability and GNP grow together is utterly startling, especially in view of the tremendous increase in military inequality among new states during the 1960's and 1970's.

So one of the assumptions of the European development model most certainly does not appear to hold for the early years of independence for these new states. There seems to be no "law of the jungle." There is no evidence that these states are arming themselves in response to external considerations; on the contrary, the size of their economies and their relationships with the Soviet Union more than any other factors explain the level in any given year of their military capability. At the same time, given the total lack of a relationship in rates of change, one cannot assume a causality between economic growth and military capability.

FIVE

Growing Arsenals

To this point, we have been dealing exclusively with global patterns; the plots and charts in the preceding chapters tell us nothing about the performance of individual countries in the sample. So despite the strong positive relationship between GNP and military capability found at the country-year level, it is still entirely possible that a plot of any given state's scores would show only a weak or even a negative relationship. In other words, plotting just the points for a single state across its years of independence could show that its military capability rose as its GNP fell, even though the general locations of its points lay on the main band of the data (this is in fact the case for one state, South Yemen).

Figure 5.1 illustrates this point. Note that though the general pattern of country-year points shows a strong positive correlation between GNP and capability, the five states whose annual scores are mapped in the data cloud show a wide range of different relationships: A shows a positive relationship between GNP and capability, C a negative one, and the others a variety of alternatives. In fact, without actually labeling the years it is not possible, even in the case of A, to say whether GNP and capability grew together or fell together.

To finally pin down national behavior, we must analyze what has gone on in the separate states, the task to which we now turn. It is not practical to trace the histories of all 46 states on a single graph, if only because their lines cross and re-cross scores of times. But it is possible to analyze state behavior both collectively and individually, and what emerges is a picture that is considerably more complex than the weakly monotonic data cloud of the country-year analysis.

National Patterns

Table 5.1 lists the correlations of GNP with military capability for 43 of the 46 states in the system (reliable GNP data were not available for

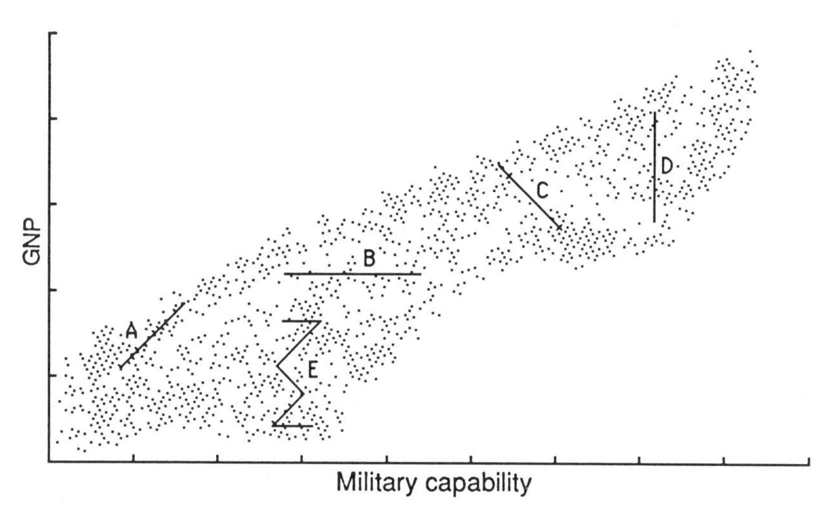

Fig. 5.1. Hypothetical country-year plot of GNP and military capability identifying scores for states A-E

the former Portuguese colonies over a long enough period to calculate meaningful averages). These numbers are a measure of how close the data points for each state fall on a single line for that state. In Figure 5.1, note that the data points for state A are not all on the line shown but rather cluster around it. The numbers in Table 5.1 show how tight the clusters are—or how close the relationship is in each state between its GNP and its military capability. The average correlation for all 43 states is 0.823, which is considerably higher than the country-year average (for the whole group) of 0.625. Moreover, the correlation scores are well clustered at the upper end of the scale; fully two-thirds of the 43 scores are above 0.8 and 86 percent are above 0.7. In most new states, therefore, military capability and GNP are quite closely related, at least when taken over the entire time span of the study.

Some of the cases where they are not are readily explained. In Niger, which displays by far the weakest correlation (0.312), the GNP fell during a severe drought from 1971 to 1975, while the country's relatively low level of military capability remained unchanged. Chad, also severely affected by the drought, had a weaker-than-average correlation (0.707), but Niger suffered a worse GNP collapse, and its correlation score was correspondingly lower. There were French garrisons in both states throughout the time period, and French military support continued during the drought, helping to keep the local military forces operating.

TABLE 5.1
Correlation Between Military Capability and GNP by Country

State	Correlation	State	Correlation
Algeria	0.7259	Malaysia	0.8978
Bangladesh	0.8395	Mali	0.7593
Benin	0.8550	Mauritania	0.9253
Botswana	0.5955	Mauritius	0.8725
Burundi	0.6467	Niger	0.3119
Cameroon	0.8640	Nigeria	0.8408
Central African Republic	0.8494	Papua New Guinea	0.9375
Chad	0.7066	Rwanda	0.8959
Congo	0.8898	Senegal	0.8803
Cyprus	0.6669	Sierra Leone	0.7335
Fiji	0.6855	Singapore	0.9560
Gabon	0.9413	Somalia	0.8757
Ghana	0.8650	South Yemen	0.5127
Guinea	0.9716	Surinam	0.9406
Guyana	0.8248	Tanzania	0.9558
Ivory Coast	0.9662	Togo	0.7673
Jamaica	0.7517	Trinidad and Tobago	0.7925
Kenya	0.9792	Uganda	0.9299
Kuwait	0.9016	Upper Volta	0.8607
Lesotho	0.8964	Zaire	0.8902
Malagasy	0.7534	Zambia	0.8751
Malawi	0.8204		

NOTE: In this table and the next, I exclude Angola, Guinea-Bissau, and Mozambique, on which reliable GNP data were not available over enough of the period.

The next-lowest correlation belongs to South Yemen, which went through an even longer period of decline than Niger; in its first ten years of independence, its GNP fell by more than 20 percent. South Yemen received substantial Soviet military assistance in the immediate post-independence period, however, and the acquisition of military equipment was clearly unrelated to any ability to pay for it. In this case, it is clear that patron support was the principal factor in a military buildup, not an increase in the resources available to fund such activity. South Yemen is somewhat atypical among the states in which a patron-shift occurred for several reasons. For one thing, it achieved independence later than many of the other states (1967), and for another, it received massive amounts of weapons from the USSR *immediately* on attaining statehood. Finally, it experienced some degree of domestic instability, which coupled with the lack of any continued investment by Britain (the former metropole), contributed to the declining GNP. Although this exact combination of circumstances was uncommon, such data as are available suggest that a similar combination of heavy Soviet military assistance in the face of declining post-independence GNP has taken place in Angola and to a lesser extent the other former Portuguese possessions in Africa.

Certain states show much lower correlations between capability and GNP over short periods of time than they do overall. For instance, in the Caribbean region Guyana and Trinidad-Tobago both showed correlations of less than 0.3 between 1965 and 1975. In Trinidad-Tobago, this period saw the retirement of some old equipment (guard boats) before all the new replacements had arrived, resulting in a temporary drop in capability while GNP continued its gradual rise; in Guyana a temporary manpower reduction produced the same result. Because of the small size of the forces involved, such temporary setbacks show up in the correlation figures because GNP growth continued unabated in each case. But they were of little significance in either the domestic or the international politics of the region.

Another example is Mali, where Soviet aid produced strong growth in military capability during the years 1960-67 despite weaker-than-average GNP growth. The country then suffered some reductions in capability because equipment was not replaced at a break-even pace while GNP stagnated. As a consequence, although Mali had a very low correlation between GNP and capability between 1960 and 1967, it grew to +0.76 for the entire period. Guinea, another Soviet client in the region, displays a much stronger correlation (0.97) than Mali, but Guinea had a much more consistent relationship with the Soviet Union and was kept more generously supplied over the years 1967-81.

There is no obvious explanation at this point for what little variation in correlation strength there is across the rest of the group, or any general explanation for why the figures vary so much when calculated over shorter time spans than the entire period. Weak states and strong states are spread out across the list, as are the rich and the poor. The strength of the relationship between capability and GNP does not vary with the size of the military force or with the size of the GNP: the actual correlation between the strength of the relationship and mean GNP is only 0.129; for mean military capability, it is hardly larger—0.197. Note that the three wealthiest states, Kuwait, Nigeria, and Algeria, are spread out all across the range, from 0.726 to 0.902. Likewise the states that underwent a patron shift are spread from 0.513 to 0.972, with the group that got increased French assistance scattered throughout that range. The distribution of correlations within the group at the upper end of the range seems largely random.

An examination of the rates of change at the country level confirms that there is little consistency between rates of change when calculated by country across the group. In absolute values, the individual national scores were well clustered at the high end, but here the average correlation by country for rates of change in military capability and GNP is close to zero, not only for annual rates of change but for the five-year averages as well. There is little pattern evident in these figures either; individual correlations range from .99 to −.98 and are spread across the range. Moreover, there is little difference in the performances of the two measures; sometimes the five-year averages work better than the annual rates and sometimes they do not.

At the state level, therefore, the findings are congruent with those of the country-year analysis. The rate-of-change relationships are too erratic to be meaningful, though there is a generally strong relationship between the annual military capability score and GNP. There are a few outliers, but their behavior is readily explained, and if the rest of the states differ in the strength of their correlations, they are all so well clustered at the high end of the scale that one can say with confidence that military capability and GNP are strongly related in new states. The final (and most important) step in the analysis is to pin down just exactly what these strong relationships are.

The Slopes

At first glance, the fact that there is no pattern in the distribution of individual state correlation scores might seem counter-intuitive. For instance, how do the states responsible for the lower right quadrant of the country-year scatter plots (those that are poor but well armed) show a high capability-GNP correlation? Remember, however, that these state-by-state correlation figures say nothing about the absolute rate of capability and GNP growth; or in terms of Figure 5.1, the degree to which a state's points cluster about its line says nothing about the slope of that line. It is quite possible (and indeed often the case) for a state to have both a steady but slow growth in GNP and (with a patron's help) a steady but rapid growth in military capability. Such a state shows a high degree of correlation between GNP and capability while at the same time winding up well armed but poor.

Regressing GNP and military capability for each state yields a figure representing not only how tightly the data cluster about a line, but also the slope (and intercept) of that line. Since we have already discovered that (with only a few exceptions) the points are tightly clustered about individual lines, we can have confidence that the slopes of those lines are meaningful statistics, and we can now concentrate on the slopes, which tell us how military capability and GNP change together in new states.

Plotting GNP on the Y axis and military capability on the X axis results in a slope that is greater as GNP rises in proportion to military capability. In other words, a steep slope (one represented by a high number) is one in which large increases in GNP are accompanied by only modest rises in military capability, and a gentle slope (low number) indicates large increases in military capability per unit increase in GNP.

Regressing GNP on military capability for the 43 states on which sufficient data are available yields the distribution of slopes shown in Figure 5.2. Notice that although the correlation coefficients were well bunched together, the slopes all spread out. One is actually negative, a large number are quite shallow, about a third of the states could be described as in the middle, and a few show very much steeper slopes than the rest. This means that the nature of the GNP-military capability relationship varies considerably across new states, and this is a critical factor in attempts to apply the European development model to them.

Table 5.2 lists the GNP-military capability slopes for each of the 43 countries, and some of the results are quite surprising. South Yemen, at the bottom, has a fairly steep negative slope; as we have seen, it is the only country whose individual performance is opposite from that of

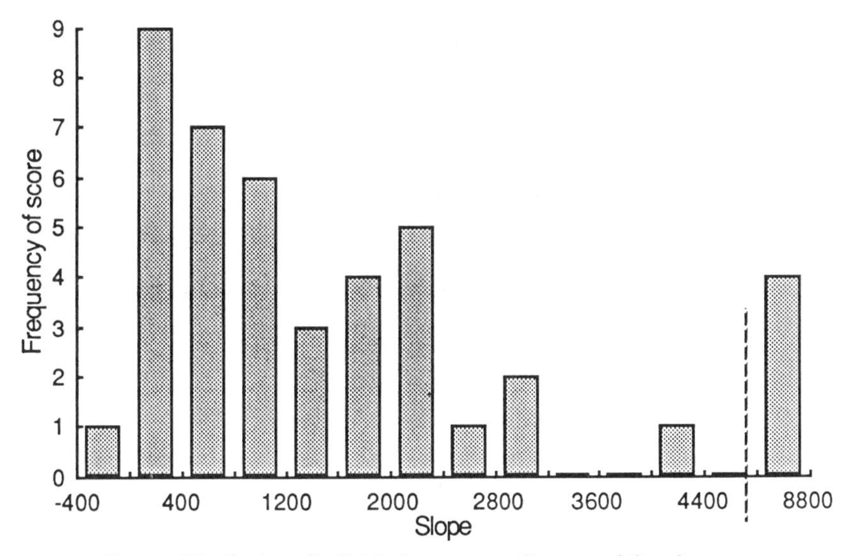

Fig. 5.2. Distribution of individual state GNP-military capability slopes

TABLE 5.2
Slope of the Military Capability–GNP Relationship by Country

State	Slope	State	Slope
Algeria	5866	Malaysia	3091
Bangladesh	2041	Mali	404
Benin	2020	Mauritania	375
Botswana	3021	Mauritius	2273
Burundi	1630	Niger	1094
Cameroon	1392	Nigeria	8504
Central African Republic	109	Papua New Guinea	350
Chad	85	Rwanda	890
Congo	327	Senegal	698
Cyprus	2161	Sierra Leone	880
Fiji	7682	Singapore	1754
Gabon	1766	Somalia	65
Ghana	861	South Yemen	−406
Guinea	303	Surinam	267
Guyana	552	Tanzania	781
Ivory Coast	2044	Togo	974
Jamaica	6567	Trinidad and Tobago	4256
Kenya	1692	Uganda	422
Kuwait	385	Upper Volta	713
Lesotho	2637	Zaire	923
Malagasy	1479	Zambia	493
Malawi	1409		

the general data cloud from which its points were drawn. At the top of the list are some of both the most populous and the least populous countries in the group, with two of the three wealthy oil states present (Nigeria and Algeria), but the third missing; Kuwait is three-quarters of the way to the bottom. Those with the highest levels of military capability are likewise well spread out, with half the top 20 percent having a steeper-than-average slope and the other half a shallower-than-average one. Interestingly, there is essentially no relationship between the steepness of the slope and the strength of the correlation between GNP and capability; the two figures correlate with each other at only a −0.16 rate.

Much of the apparent randomness in this list can be resolved into a coherent pattern with a more rigorous statistical analysis. It is possible to correlate the value of the slopes with indexes of performance on development and military capability. In other words, one can precisely calculate whether the steepness of the slopes varies with how rich or powerful a country is or the rates at which it grows on those dimensions. A summary of ten such analyses is given in Table 5.3.

The first thing that is striking about the numbers displayed in Table

5.3 is how little performance in military capability acquisition has to do with the slope of the GNP-capability graph. There is a low correlation between capability and slope, showing that, to a slight extent, states scoring high on military capability have a steeper slope than those scoring low. There is even less of a correlation between capability and rate of growth, but when rate of growth is divided by capability (this controls for the size of the force and essentially expresses growth rate as a percentage), the correlation grows a bit stronger, and importantly, this correlation is negative. This means that to a certain extent (not a lot—r = 0.2 out of a possible 1.0—but measurably) countries that grow the fastest in military capability have a shallow slope; they tend to translate limited GNP growth into large capability increases. This finding does not support a feedback model, but suggests instead either that countries whose capability grows quickly achieve this at the expense of GNP growth or that patron support goes most heavily to countries that do poorly in GNP. Further digging in the individual country data is required to resolve this issue.

There is almost no correlation between the stability of the growth rate and slope, but the highest correlation among the capability statistics shows up between slope and the stability of the capability score itself. Moreover, it is a negative correlation, suggesting that those states with the most unstable capability scores (highest standard deviations) tend to be the ones with the shallow slopes. Why this should be so is not immediately obvious, but the explanation again lies in the role of patron support, which tends to produce large increases over

TABLE 5.3
Correlation of GNP–Military Capability Slopes with GNP and Capability Statistics

Statistic	Correlation
Capability	
Military capability	0.1949
Military capability growth rate	−0.1180
Growth rate divided by capability	−0.2002
Standard deviation in capability	−0.3071
Standard deviation in growth rate	−0.0888
Gross national product	
GNP	0.5468
GNP growth rate	0.5737
Growth rate divided by GNP	0.3713
Standard deviation in GNP	0.6427
Standard deviation in growth rate	0.4586

time in military capability and thereby raise the standard deviation in that statistic when calculated across the period. Therefore, those states whose capability is most dependent on patronage have higher standard deviations in capability as well as shallower slopes.

Conclusions drawn from any of the capability statistics should be tempered with the reminder that the correlations, although interesting, are in each case quite low. The development correlations are much stronger, however, and the bottom half of Table 5.3 offers some fascinating insights. All the GNP statistics are positive, and most are more than twice as strong as their capability counterparts. Performance on GNP appears to go hand in hand with a steep slope; in other words, states that do well in economic growth tend to translate that growth into military capability to a lesser extent than those that do poorly.

This is a strong indication at the country level that as of now there is no feedback relationship between development and capability in new states. If there were such a relationship, one would expect those states that were most successful in acquiring military capability to do best in GNP growth, and conversely, those that were most successful in economic growth to be those whose military capability needs drove them. The opposite appears to be the case. The higher the GNP or its growth rate, the steeper the slope; those countries that are most economically successful get less military capability per unit of GNP than those that do not do as well.

The standard deviations tell the same story. The best performers have a high standard deviation in GNP because their absolute progress on this indicator is the highest across the whole period. The correlation between GNP standard deviation and slope is, in fact, the highest of any in the table, showing that the best performers have the steepest slopes. There is also a fairly strong correlation with standard deviation in rate of change, mostly because those states with the highest growth rates tend to have less stable rates of change. For this group, even the best performers depend largely on commodities for their GNP, and those with a lot to sell have erratic figures when markets are unsettled.

In conclusion, the aggregate picture is becoming clear. In general, military capability grows with GNP (and in one case it grows without it), but the countries doing the best economically tend to do proportionately the least well in military capability acquisition. However, the statistics leave plenty of room for variation among the group, and they do not, by themselves, pinpoint completely the role of patrons or any other intervening factors that might affect the relationship between development and military capability. We must therefore look more closely at some individual cases to see if we can detect the mechanisms that are generating these figures and find the explanations behind the very interesting aggregate findings.

Patterns of Diversity

We start with Ghana, which is a country of average size and wealth and exhibits many of the patterns found throughout the group. Its annual GNP and military capability scores, with each point identified by year, are shown in Figure 5.3. Also shown is the regression line, giving a visual indication of the slope and the degree to which the points cluster around it. Ghana's slope is slightly shallower than average (26th of 43), and its data points are clustered slightly more tightly than average (20th), making it a good middle-range state with which to begin the analysis.

As we see, military capability and GNP both proceed generally upward in Ghana until after 1963, when GNP growth begins to falter. Over the next six years GNP essentially stagnates, but growth in military capability just gradually slows down, finally stopping only in 1968. It does not keep even with the GNP score, but continues to grow

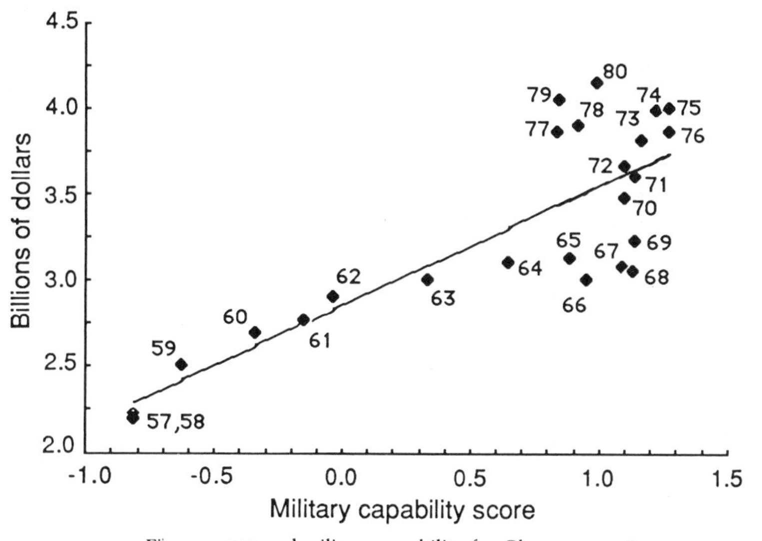

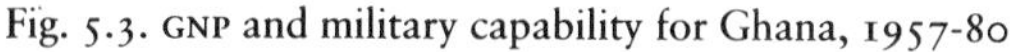
Fig. 5.3. GNP and military capability for Ghana, 1957-80

in the face of poor performance and in the end does not fall but simply remains constant. As noted in Chapter 2, as the first new state in Africa, Ghana received military assistance from a great many nations; and though by the early 1960's President Nkrumah's leftist politics had begun to alienate Western supporters, the Soviet Union provided increased aid through this period, taking up the slack and helping to account for the continued military growth in a period of economic stagnation.

By the time GNP growth resumed, most of Ghana's weapons were nearly 20 years old and in need of replacement. The effects of the re-equipment period, which applied in other states as well, were an element in the confusion in the rate-of-change data discussed in Chapter 4. Nkrumah was ousted in a coup in 1966, and his successors were conservative military officers who distanced themselves from the Soviet Union and thus from its military largess. After a few years of good GNP growth, the economy faltered again, and the re-equipment cycle, which had been holding capability constant, could not be completed with the internal resources available. The removal without replacement of some very old aircraft in 1977 reduced the military capability score, which recovered only very slowly in the lean economic climate.

Reasons why military capability does not fall during bad economic times are not hard to find. Not only does the military have the bureaucratic and coercive power to somewhat insulate itself from times of general economic hardship, but capability, as measured by weapons stocks, is resistant to budget austerity. Military weapons have limited resale value and the money gained by selling off weapons stocks during lean times is minimal compared to the costs incurred in replacing them later. So, although cutbacks may force reduced operating schedules and limit procurement, any loss of operational effectiveness is masked because existing weapons stocks are almost always left untouched. As a consequence, there is a ratchet effect; when times are bad or economic prospects unpredictable, a state's military capability may stagnate but is almost never reduced. We can see this general pattern by looking once again at Figure 4.10 (p. 75). Note that the rate of change in military capability almost never goes below zero, although the rate of change in GNP often does so. Military capability, in fact, declines in less than 2 percent of the country-year cases for which a score was calculated. The ratchet effect is a general phenomenon. Military capability moves in only one direction—up—whereas the development index, although the long-term trend is positive, can go either way in any given year. Those reductions in military capability that are observed generally fall in replacement cycles, which if interrupted by a funding crunch (as in the case of Ghana) can lead to an extended period of lowered scores.

As might be expected, clients of the USSR tend to have shallower slopes than most (i.e., they have a lot of military power for not much GNP growth). Six of the 12 shallowest slopes fall in this category, and as just noted with Ghana, Soviet assistance can have an impact even in cases where no coded patron shift has occurred. (The other six states are typically poor and weak, with low capability growth but even lower GNP growth.) Since the French were also responsible for some relatively poor but powerful countries, it is interesting to compare their impact on clients with that of the Soviet Union.

Figure 5.4 plots GNP and military capability for three former French possessions—the Ivory Coast and Cameroon, which remained in the French Community, and Guinea, which aligned itself with the Soviet Union. A caution, to be repeated as a preface to all subsequent figures, is to keep in mind where each country lies on the scatter plot of the full data. The GNP range shown in Figure 5.4 is greater than the range shown for Ghana (Ghana's begins at two billion dollars against a half billion here); many of the plots for Ghana after 1966 would be just off the upper end of Figure 5.4's capability scale.

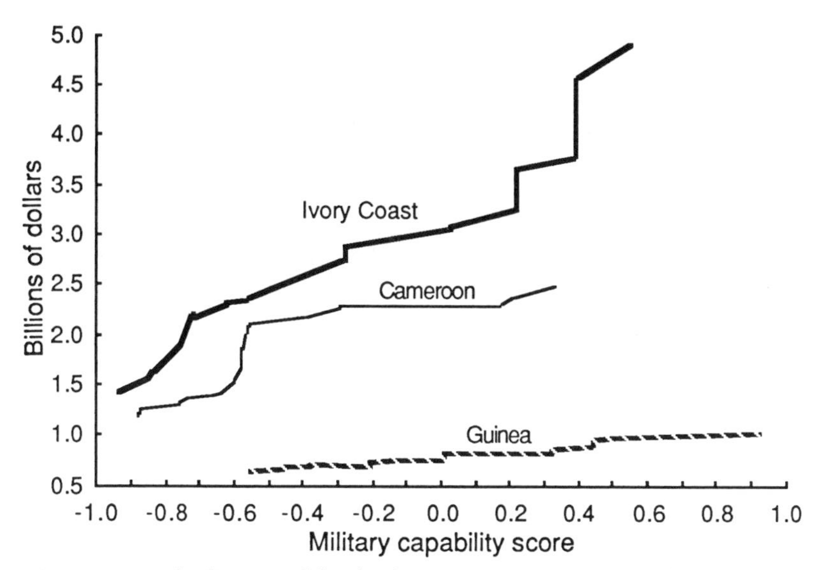

Fig. 5.4. GNP and military capability for the Ivory Coast, Cameroon, and Guinea. All three became independent in 1960.

The most striking thing about Figure 5.4 is the performance of Guinea, which easily achieved the highest military capability score of the three states while remaining by far the poorest. Guinea has one of the shallowest slopes in the data set (38th of 46). Even though it is among the poorest of all the states and has a much lower-than-average GNP growth rate, its average military capability puts it in the top quartile, and in percentage growth rate in military capability it ranks third!

Guinea's poor economic performance is probably not a direct result of its military burdens, because Soviet assistance did not cost it very much in monetary terms. Personnel dislocations (the military absorbing a significant fraction of trained and educated people) may have been a factor, but the antipathy of Western investors was certainly important.

In contrast, the Ivory Coast, which encouraged a continued French economic presence, saw a steady and sometimes spectacular growth in GNP; at the end of the period, its GNP was more than six times as large as Guinea's. Nevertheless, the Ivory Coast has never equaled Guinea in military capability, and if it continues to have a lower rate of capability growth than Guinea, it will never catch up. Cameroon has been less economically successful than the Ivory Coast, in part because the government faced the problem of integrating territories formerly under separate French and British rule. There has been some insurgency, and owing in part to the military response to it, the country has a shallower-than-average slope. The growth in military capability during a period of very flat economic performance was only possible with French assistance, but that assistance amounted to far less than the Soviet aid to Guinea.

Figure 5.4 illustrates in detail two important results of the general analyses. First, states that do best economically tend to have steeper slopes on the GNP-capability graph; they tend to acquire less military capability per unit of economic wealth than those that do poorest economically. Second, a patron's support can offer a state military capability equal to or greater than states of comparable size with much stronger economies. The lesson seems to be this: if your regime decides it needs a lot of military capability to survive, you can get it from a friend at least as easily as you can acquire it on your own.

A comparison of the arsenals that produced the capability scores in Figure 5.4 is also interesting. By the late 1970's, the Ivory Coast had a medium-sized army, air force, and navy, but possessed almost no high-capability weapons. Apart from a few modern transport planes, the air force consisted almost entirely of utility aircraft and light helicopters.

The Ivory Coast had no fighters in its air force, no tanks in its army, and only one patrol boat in its navy. In the same period, Guinea, with only slightly more military personnel (10,000 versus the Ivory Coast's 8,000), had nearly two dozen MIG fighters, more than 80 tanks and armored personnel carriers, and about 10 high-capability patrol boats.

The differences highlighted here are typical of the French and Soviet approaches to military assistance. France has consistently avoided providing high-capability weapons, but has been quite generous with others. This mirrors France's intention to protect its clients against outside threats itself, and that is the only mission for which high-capability weapons are required. The French are not interested in provoking any arms races between states that they themselves will have to fuel, nor are they interested in making their own position more difficult by arming too well states that might turn on them. Still, the French value their relationship with their former colonies enough to equip them generally better (controlling for size and GNP) than the states that were formerly British possessions.

The Soviet Union, on the other hand, has had limited military intervention capability and no interest or experience in balancing regional competition. Therefore, it equips all of its clients like little Soviet Unions—lots of armor, jet fighters, and torpedo-armed patrol boats. The military forces of Soviet clients are not always larger than others, but they are usually much better equipped and well balanced across all categories, including the most expensive and high-capability weapons.

These patterns remain remarkably consistent even for much wealthier countries. Figure 5.5 plots GNP and military capability for the three wealthiest states, Nigeria, Kuwait, and Algeria. Note that the GNP scale begins where Figure 5.4's ends and covers a range *seven* times as wide, and that the capability scale too goes much higher (to the top of its limit) and covers nearly twice the range of the earlier figure.

As before, the state receiving Soviet weapons, Algeria, scores generally better on military capability even though for most of the period its GNP is the lowest of the three. But the picture here is more complex. Nigeria shows a very interesting pattern as the result of its civil war. Generally underarmed for its wealth (compared with the other new states), Nigeria had a low capability score until the Biafran secession of 1967. Over the next three years of very severe conflict (more than a million probably died of wounds and starvation), the GNP fell by about a sixth while military capability grew enormously. This growth was essentially based on internal resources, for though the Western countries were willing to extend credit, they did not give away the hardware as in

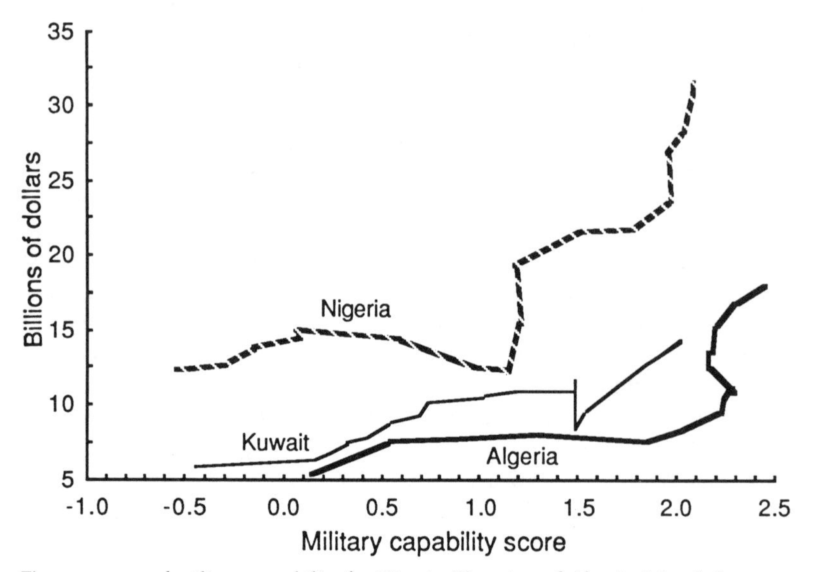

Fig. 5.5. GNP and military capability for Nigeria, Kuwait, and Algeria. Nigeria became independent in 1960, Kuwait in 1961, and Algeria in 1962.

years past. At the end of the war, there was only a very limited demobilization, so military capability did not fall, but with the resumption of oil production, GNP shot up and began climbing in about the same pattern as the one before the conflict. Subsequently, military capability and GNP again began to move together at about the prewar rate, until rising oil prices combined with production increases caused GNP to shoot up once more.

Nigeria illustrates three important points that apply to many of the countries with initially small forces. First, there are internal resources available for substantial increases in military capability, but the motivation must be present to draw on them. Over the period of the civil war, Nigeria built a military force many times more powerful than the one it had at independence despite a declining GNP. Second, when GNP grows rapidly, the military forces of new states are unable to absorb the newly available resources fast enough to maintain their share of national expenditures, so that capability often "stair-steps" upward as procurement decisions made in good economic years are implemented across later periods. Finally, to put this internally fueled capability in perspective, note that Nigeria's score at the end of the civil war was about the same as Guinea achieved on less than a tenth the GNP.

(Guinea had far fewer military personnel, but a comparable number of ships, planes, and armored fighting vehicles).

Like Nigeria, Kuwait came to independence virtually unarmed. The Kuwaitis' attention was quickly drawn to defense matters, however, when Iraq immediately challenged their sovereignty. The British dispatched military units to guarantee the new state's independence, but the lesson took. Kuwait acquired military capability at a remarkably steady rate over the next 20 years, to the point that it had the only slope in the bottom 12 belonging neither to a Soviet client nor to an exceptionally poor state. This steady growth was made possible by Kuwait's enormous per capita GNP, which essentially allowed so much slack in the budgeting process that whether oil prices were going up or down there was always more money available than the defense establishment could absorb.

Algeria, the one state among these oil producers that has had close ties to the Soviet Union since independence, achieved statehood via revolution, and began with a much higher-than-average capability score. Algeria's slope is steeper than average, since it did not have to go as far as other states to reach its very high end-level of military capability. With Soviet assistance, Algerian forces expanded throughout the first ten years of independence. Economic growth was rather flat during most of this period and did not take off until the oil price rises of 1973. Algeria has sought to diversify its suppliers since then and has had the wealth to do so. Its growth in capability slowed when the government embarked on an extensive re-equipment program in 1974 and resumed when the program was completed in the late 1970's.

Despite the obvious differences in scale and individual histories, these three states show the same pattern as the one we saw in the much smaller states of Ivory Coast, Cameroon, and Guinea. The greater the GNP growth, the steeper the slope, and the Soviet-supplied state achieved a higher capability level than states that had more wealth at their disposal but depended less on Soviet arms transfers.

A final point about the Soviet role in equipping new states can be seen in Figure 5.6, which covers Somalia, South Yemen, and Togo. Again, note the scales: the GNP scale reaches only $700 million dollars, *one-fiftieth* the height of Figure 5.5's. Were the data in Figure 5.6 plotted on the same scale as Figure 5.5, the lines would be off the bottom of the chart. Nevertheless, the capability scale is the same as Figure 5.5's and tops out at 2.5.

What we observe in Figure 5.6 is the effect of Soviet military assistance on two of the poorest new states in the world. South Yemen has

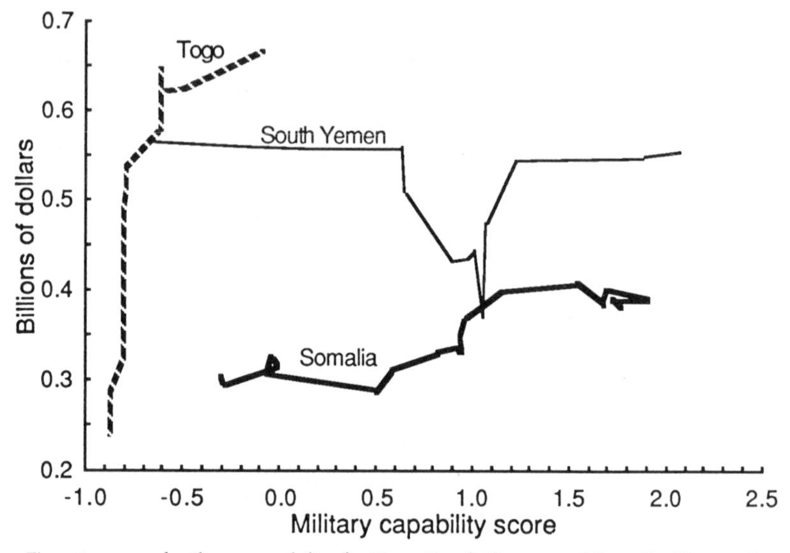

Fig 5.6. GNP and military capability for Togo, South Yemen, and Somalia. Togo and Somalia became independent in 1960, South Yemen in 1967.

had an essentially stagnant GNP ever since independence (with the exception of a six-year dip and recovery). Nevertheless, with Soviet aid its military capability score grew with extraordinary consistency, to the point where it essentially equaled that of Nigeria, whose GNP was an incredible 60 times the size of South Yemen's! For Somalia the story is similar; a slow growth of GNP matched with a rapid growth rate in military capability led to a point in the mid-1970's where a state with one of the lowest GNPs had one of the highest capability scores. Somalia lost its Soviet patron in the late 1970's, however, when the Russians switched to Ethiopia after the overthrow of Haile Selassie. Somalia's capability score then fell and never returned to its peak because equipment lost in combat with Ethiopia over the Ogaden Desert in 1977-79 was not replaced.

Although it is striking enough that South Yemen and Somalia are both in the top fifth of the states in military capability, while falling in the bottom third in GNP, more striking still is their GNP growth *rates*. South Yemen is in last place, having the only negative overall rate in the group, and Somalia is right above it. As with Guinea, however, it is hard to ascribe this poor economic performance to defense burdens because both countries received most Soviet equipment in exchange for basing rights, not money or commodities. Rather, it appears that the attention of the leadership in both states has been much more focused on military ambitions than on development.

Somali peoples live in Djibouti, Ethiopia, and Kenya, as well as Somalia, and Somali rulers have pressed irredentist claims against those countries ever since the state gained its independence. They explicitly rejected early Western offers of assistance as insufficient to carry out their planned conquests and invited Soviet support because it was viewed as *militarily* more generous. It certainly was, but by the time Somalia actually got around to invading Ethiopia, the Soviets had switched sides. With as many as 40,000 Cuban troops fighting alongside them, the Ethiopians were able to repulse the Somalis, who were unable to get any significant military assistance from the West after the Soviets left. As for South Yemen, it has had ambitions over the Dhofar region of Oman, as well as conflicts with North Yemen over their ill-defined borders. In both cases, then, the pattern expected from the European development model is reversed; when external security issues preoccupy elites, development concerns fall by the wayside.

Togo, a state whose GNP spans the same range as Somalia's and South Yemen's, is added to Figure 5.6 for a comparison. Togo is by no means a model of stability and development; it has experienced several military coups, and many states have done better on GNP. Nevertheless, the contrast with Somalia and South Yemen is striking. Togo went from being poorer than either to being richer than both during the period 1960-75, while being utterly outdistanced in military capability. To put the figure in another perspective, Togo's slope is exactly in the middle of the range (22d of 43), which gives an idea of how extremely flat the slopes of the other two are. Once again, the better the economic performance, the steeper the slope; and the completely out-of-proportion capability scores are entirely an effect of patron support.

Of course, there are new states with high military-capability scores or high GNPs that did not depend on oil for their wealth or the Soviet Union for their military power. Figure 5.7 plots the curves for three of the top six scorers on military capability: Malaysia, Singapore, and Zaire, which rank 2, 3, and 6, respectively, in average military capability (Algeria, Kuwait, and Nigeria rank 1, 4, and 5).

Zaire's GNP is fairly large for Africa, but it is only about half as high as the Ivory Coast's, a country that has a much lower capability score. Zaire, however, had a patron—the United States. In the turmoil that surrounded its break from Belgium in the 1960's, Zaire came to have an important symbolic value to the United States, and thanks to American aid, much of it in the form of unpublicized CIA technical as-

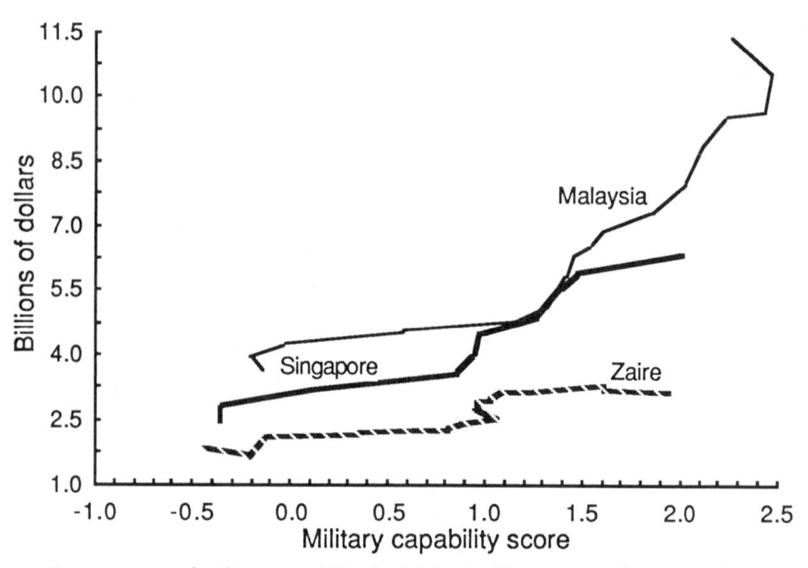

Fig. 5.7. GNP and military capability for Malaysia, Singapore, and Zaire. Malaysia became independent in 1957, Singapore in 1965, and Zaire in 1960.

sistance, the country's military capability grew very substantially. Like the Soviet Union, however, the United States had little experience in Africa, and Zaire wound up with a force that bore a resemblance to its patron's. It had lots of air transport and helicopters, jet fighters but no anti-aircraft missiles, and light patrol craft with no torpedoes or anti-ship missiles, all typical of the U.S. forces. The French also were heavily involved in Zaire, and this tended to limit U.S. influence over the long run. Most of the army used French equipment, for example, and so tended to follow the French practice in Africa of favoring armored cars and light APCs over tanks.

Although U.S. assistance to Zaire was on a less massive scale, proportionately, than the amounts received by Soviet clients, one can see some of the same effects. Zaire's forces are structured quite differently from those of its neighbors; they are not outsized for the area and the population, and are equipped with generally more advanced weapons than other states in the region. Zaire's slope is shallower than average, and it ranks low (30th of 43) in the rate of GNP growth.

The other two states in Figure 5.7 have done far better economically than Zaire, but, once again, note the GNP scale, which reaches only about a third as far as the scale in Figure 5.5 (the wealthy oil producers). Malaysia exceeded Nigeria and equaled Algeria in military capability, with a smaller GNP and no patron for much of the time. Regional military threats account for this high score. For most of the early 1960's Malaysia was harried by Indonesian forces trying to seize the Malaysian states on the island of Borneo. This attempt was thwarted with British assistance and provoked a large increase in capability during a period of slow economic growth.

Singapore, which accounted for about half the GNP of the Malaysian Federation, broke away from the federation in 1965 for reasons having to do with ethnic hostility between the Chinese and Malay communities. The military was almost entirely Malay, and Singapore gained independence with virtually no defense forces. Initially content to let the British provide for its defense in exchange for basing rights, Singapore began a concerted military buildup with the British Labour government's decision to withdraw from areas east of Suez by 1971. The buildup was not aimed at Malaysia—indeed, the regional Commonwealth states (which include Australia and New Zealand) closely coordinate defense planning—but with the turmoil in Southeast Asia and the departure of the United States from Vietnam, the complete lack of defense forces worried the country's leaders.

Singapore's economy grew rapidly in the 1970's (by many measures it should now be considered a developed state), and with the help of some oil revenues from Borneo, so did Malaysia's. Their military growth kept pace. Singapore's lagged a bit because of its small size; as with Kuwait, there were absolute limits on how far it could grow. The dip in Malaysia's capability in the last part of the period was due to a reduction in the size of the army. That decision appears to have been reversed in 1981. Both Singapore and Malaysia have steeper-than-average slopes, though not as steep as Nigeria's and Algeria's; in those states oil revenue simply arrived too fast for the military to spend.

What this trio of states illustrates is that the relationship between economic performance and slope applies to new states at the upper end of the development range, and that high levels of capability can be induced by patrons other than the Soviet Union. These states are much wealthier and more powerful than the average new state, however, and a glance at the other end of the scale is instructive. Figure 5.8 plots GNP and military capability for Sierra Leone, Rwanda, and Fiji, all of which are in the bottom quartile in military capability. The GNP scale here is identical to the scale in Figure 5.6 (Somalia, South Yemen, and Togo), but the military-capability scale starts lower and goes only *one-seventh* as high (up to only the lowest score for Somalia).

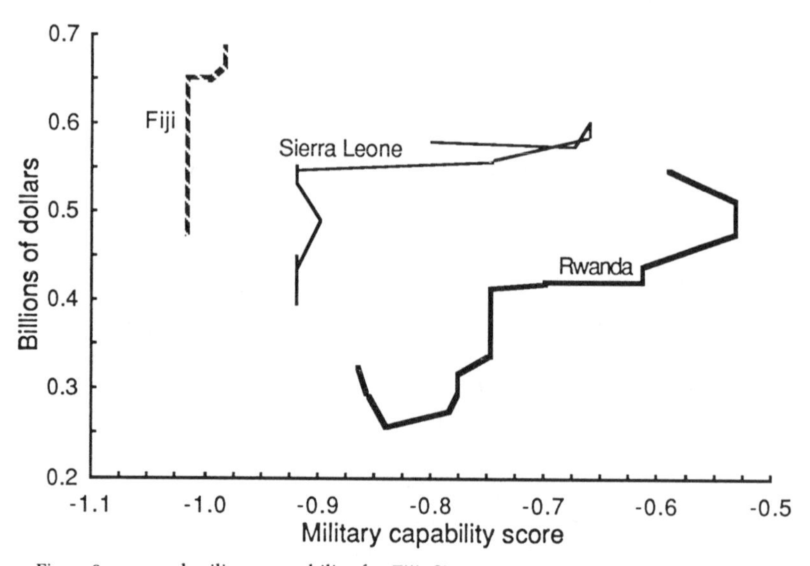

Fig. 5.8. GNP and military capability for Fiji, Sierra Leone, and Rwanda. Fiji became independent in 1970, Sierra Leone in 1961, and Rwanda in 1962.

Fiji is probably as isolated from external threats as any new state could be. Not only is it an island chain, but it is remote from areas of contention and has no neighbors with significant military forces. Fiji has the lowest average military capability score in the data set, an interesting point since it frequently rents its army to the United Nations as a peacekeeping force and counts on this use of its troops to earn foreign exchange. The potential for growth of this "business" is restricted, however, because of the divisive ethnic politics of the island. Only 42 percent of Fijians are of the original Melanesian stock; most of the others are Indians, descended from the indentured plantation workers brought in by the British 80 years ago. Although a minority, the Melanesians are guaranteed most of the land and make up most of the army, but as long as the army remains small, it has limited utility as a potential agent of Melanesian domination. Therefore, partly in the interests of ethnic peace, the army has been kept down to about 700 men, which is smaller than average for a country of nearly 600,000 citizens.

Fiji has the steepest slope in the data set; indeed, for most of its existence as a state, the slope has been vertical, because no expansion of military capability took place. The slopes of the other two states in Figure 5.8 are about the average (24th and 25th of 43), although Sierra Leone ranks a low 39th in average military capability, and Rwanda 31st. To put their capability in perspective, note that their GNPs are in the same range as South Yemen's and Somalia's, but that they never achieve capability scores as high as the lowest-year score for Somalia. In fact, if plotted on the same scale, the countries of Figure 5.6 would require a figure six times as wide.

Rwanda and Sierra Leone illustrate some points common to low scorers in military capability. First, they show a fall in capability more often than states with larger forces. This is because equipment wear and replacement cycles hit them very hard: in small forces, the impact of removing even a few weapons from the total is enough to send the score lower, unless replacements arrive at the same time. Second, growth occurs sporadically, because the budget is not large enough to provide for continuing purchases. And finally, growth often occurs after a few good years of GNP growth and coasts upward for a while after such growth ceases. Unless there is an external conflict or a patron providing assistance, growth in military capability almost never precedes GNP growth. Capability growth usually follows GNP growth, a pattern that can be seen in all three states here.

Neither Rwanda nor Sierra Leone had especially stable governments in their first 20 years. There were a number of military coups in Sierra Leone, but on assuming power, none of the military rulers made any great effort to increase the country's military capability. On the contrary, there was a substantial reduction of capability under military rule in the mid-1970's. Rwanda was severely shaken by ethnic turmoil in which hundreds of thousands of people died or were driven into exile. Again, however, the military, which was involved in the bloodshed, did not use its power to acquire a much larger and more sophisticated weapons stockpile.

Figure 5.8 illustrates that in the absence of patron interest or external threat, capability will tend to remain low when GNP remains low. This is the case even when there is considerable domestic conflict in which the military is involved. As with much larger states, the pattern holds that the better the GNP growth rate, the steeper the slope of the GNP-capability plot, although the lines get relatively erratic when forces and economies as small as these are involved.

A final set of examples is shown in Figure 5.9, which plots GNP and military capability for Kenya, Uganda, and Tanzania. These three states are an interesting set, because they are all in the same region, are roughly the same size, and share a common colonial heritage. Despite

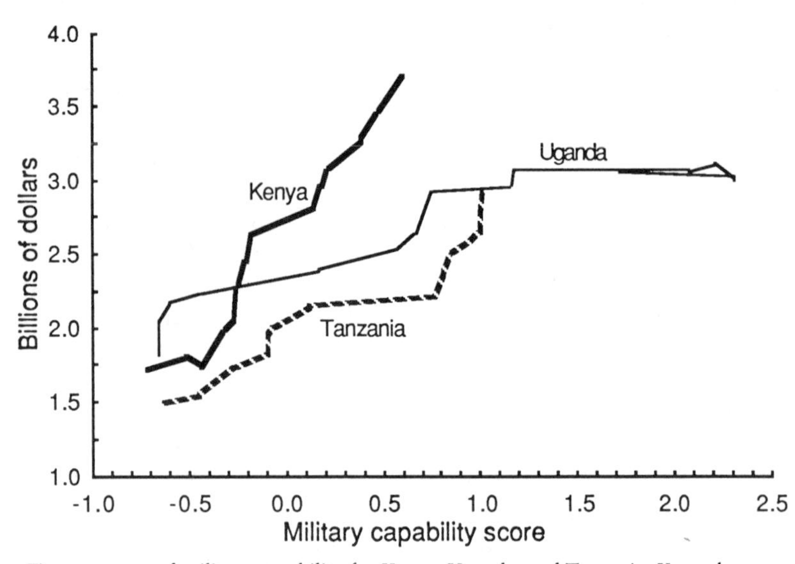

Fig. 5.9. GNP and military capability for Kenya, Uganda, and Tanzania. Kenya became independent in 1963, Uganda in 1962, and Tanzania in 1964.

these similarities, however, they have had very different domestic political experiences, and each took a quite different path in its foreign policy and security relationships.

Uganda has had one of the saddest post-independence histories of all the new states. As we have seen, its army was created from a battalion of the King's African Rifles that had seen action in both world wars and in suppressing the Mau Mau insurgency in Kenya during the 1950's. Following British practice, the army was recruited largely from the Moslem peoples of the north, although at independence in 1962 the Christian Buganda tribe was dominant politically. Like the armies of the other two former British East African countries, Uganda's army mutinied in early 1964 over pay and the continued presence of British officers. At the invitation of the government, the British intervened to put down the mutiny, and although his role in the affair was ambiguous, Major Idi Amin emerged as one of the new battalion commanders after the British left. His battalion led an assault on the presidential residence in 1966, helping the prime minister, Milton Obote, to consolidate his power by driving the Bugandan president from the country. Then, in 1971, Amin, by now the army commander, took power himself, staging a successful coup against Obote when he was out of the country.

Before Amin came to power, Uganda had received military equipment and training from many sources, including Israel, the Soviet Union, and Britain. After the 1971 coup, Amin again sought aid from Britain, but when it became clear that the British would give him only limited assistance and would expect him to pay for most of the hardware supplied, he turned elsewhere. Having expelled the Israelis, he appealed to Libya, which offered personnel and equipment, and to the Soviet Union. Five years later, thanks to Soviet aid, Uganda had one of the best-equipped armies in Africa.

Note that the military capability scale in Figure 5.9 covers the same range as Figures 5.5-5.7, but that the GNP scale begins above the scale for Somalia, South Yemen, and Togo (Fig. 5.6), falls below the scale for Nigeria, Algeria, and Kuwait (5.5), and rises only a third as high as the scale for Malaysia, Singapore, and Zaire (5.7). Uganda's tumultuous military history can be easily traced in Figure 5.9. After a couple of years of no growth, the British officers left, and military capability rose steadily until after 1971, when it shot up very rapidly. The growth reversed after 1976, however, because the Soviet Union, acknowledging the increasingly erratic behavior of Amin, did not fully replace the aircraft lost in the Israeli hostage rescue at Entebbe.

Although Tanzania too has had some turmoil, it has had a much more stable history than Uganda. It went through a mutiny in 1964, which was put down by the British. At about this time, Tanzania also incorporated the island of Zanzibar, shaken by disorder after the Arab-dominated regime was ousted in a Black-led coup. Tanzania chose to follow the Chinese and Soviet approach, viewing the army as part of the political apparatus, subject to the control of the ruling party and a target for political indoctrination. Furthermore, Tanzania achieved closer military ties with China than any other African state, although it also obtained military equipment from Soviet and Western sources. As a result of its ties with the Soviet Union and China, Tanzania has a shallower-than-average slope, though not so shallow as Uganda's (27th of 43 versus 32d for Uganda).

Idi Amin's often-proclaimed territorial ambitions included large pieces of Tanzania (as well as Kenya), resulting in border clashes as early as 1972. In 1979, the Tanzanians found the situation so intolerable that they invaded Uganda, hoping not only to end the disorders on the border, but also to eliminate the Amin regime. In both respects they were successful, for Ugandan resistance quickly collapsed, despite a halfhearted intervention by Libya.

The result of this clash points up a limitation of the military-capability index as discussed in Chapter 4. According to the index, at the time of the Tanzanian invasion, Uganda had more military capability than its attacker. The raw numbers on weapons stocks would seem to confirm this. Uganda had several times as many tanks and armored personnel carriers as Tanzania, and theoretically their air forces were nearly equal in size and sophistication. Nevertheless, the issue was never close. Much of Uganda's equipment was not operational, most foreign advisers had fled the chaos of Amin's last years, and the army was undisciplined and poorly led. Consequently, although the capability score for Uganda shows a decline after 1976, the decline does not fully capture the extent of the general military collapse that took place.

In contrast to Uganda and Tanzania, Kenya had a consistently higher GNP growth rate and a generally slower military buildup (its slope was much steeper—16th of 43), even though Kenya had some of the same problems as Uganda at independence. The majority Kikuyu tribe made up only a small part of the army, the result of British policy and of the effort to suppress the Mau Mau, who were primarily Kikuyu. There were also mutinies in 1964 as in the other two states. Three factors combined, however, to aid Kenya in professionalizing its military. One was that its somewhat later independence (1963) left the early experiences of Uganda and Tanzania as examples. Another was the personal prestige of President Jomo Kenyatta, which enabled him to proceed with an ethnic integration of the military that was not undertaken in Uganda. Kenyatta also maintained a close relationship with the British, thereby ensuring continued help in officer training and equipment maintenance all through the 1960's and early 1970's. Finally, Somali claims against Kenya's northern territory had to be taken seriously from the first (recall that Somalia built one of the largest military forces in Africa), and so there was an obvious need for military preparedness.

Nevertheless, Kenya witnessed several attempts on the part of individual senior officers to play political roles, which raised tensions and resulted in their dismissal; and an attempted coup by the air force against Kenyatta's successor was put down only with violence. It is not yet clear which approach, in the long run, will be more successful in containing military coups—Kenyan professionalism or Tanzanian politicization. Julius Nyerere so thoroughly dominated the politics of Tanzania that it is hard to foresee what will happen now that he has finally retired. And in the case of Kenya, although it has clearly done better by most measures than its neighbors, attempts to portray it as a triumph of Western ideals in Africa ignore not only the coup attempts, but also the fact that in per capita terms the growing national GNP has hardly stayed even.

In most new states, military capability has been so low that it has been irrelevant outside the domestic context. In these three countries, however, military power reached the point that it had international implications (i.e., they could invade each other). But even then, such a capability did not alter the pattern found elsewhere. Steeper slopes went with better rates of economic development, and when a leader decided a large increase in military capability was important, it was obtained from patrons.

Summary

Despite the great variation in individual state performance in economic growth and the acquisition of military power shown in these examples, it is possible to discern some general patterns. Military power tends to increase in the wake of economic progress rather than precede it, unless the military power comes as the result of a patron's support. And of all the patrons of new states, the Soviet Union is the most generous when it comes to military assistance. Patron support can easily provide capability scores equal to the highest found in the data set, and so regimes interested in military power have an alternative to domestic resources in acquiring it.

In general, those states that did best in GNP growth appear to have paid less attention to military capability than others. This relationship is not perfect, but holds right across the range from poor states to rich and from weak states to powerful. Those that did most poorly in GNP growth not only paid more attention to military capability, but usually chose a patron's aid as the route to acquiring it. Whether the regime was military or civil, or came to power peacefully or via a revolution, had no obvious impact on the GNP-military capability relationship. In the final chapter we will put all this together and examine the not altogether pleasant implications that these findings hold for development and conflict in new states.

SIX

Patterns and Trends

The three preceding number-filled chapters have provided various pieces to a picture of the military's role in the development of new states. It is now time to fit those pieces together in a way that will make the data useful not only as an explanation of the past, but also as a window on the future.

Our purpose has been to examine whether the European model of development applies to the new states of our time. This model postulates, first, that development is an arduous and even dangerous process that elites undertake only in response to external threats they cannot meet in any other way, and second, that for all the difficulties and dangers, successful development results in states whose citizens are ultimately much better off than they were before the process began. In fact, this European model does not fit the behavior of modern new states, and there is little prospect that it will in the forseeable future. The findings of this research at all levels of analysis fit together into a picture that leaves little room for ambiguity about this conclusion and, unfortunately, much room for pessimism about the development prospects in a large part of the newly independent world.

Findings

In the European model, as we have seen, there is a feedback relationship between development and military power, with the need for military power driving development, while development itself provides the wherewithal for the creation of military power. When such a model is operating, it should be possible to observe some very straightforward effects. One would expect, for example, that progress in development and military capability would be correlated, and that the closer the correlation, the more effective the operation of the feedback process.

One would also expect that development would proceed most quickly when there was a need for military power, and that this could be observed either by comparing development-military capability relationships (slopes) or by comparing development to the external security environment.

None of these effects are observed for modern new states. At the individual state level, GNP is correlated generally strongly with military capability, but such variation in the correlations as exists is not related to development. The strength of the GNP-military capability relationship does not go up with either the size, wealth, or military power of a state or with its growth rate in wealth or military capability. Those states that do best on development or military capability do not necessarily demonstrate the tightest relationships between these two variables. The pattern is random; each score can be explained on an idiosyncratic basis, but there is no one explanation for the overall distribution. If there is any evidence in the correlation figures for a feedback mechanism operating differentially across new states, it is so weak that it is lost in the noise.

On the other hand, GNP and military capability correlate very highly across all states, a finding that is to be expected from the model. Since the correlations are high, the lines produced by regressing GNP and military capability can be a meaningful focus of analysis. These lines represent the actual relationships between the two variables and show most clearly the mechanisms operating. There is a lot of variation in the slopes of these lines; they go from negative to highly positive with a broad band in between, so they offer the best possibility for observing any feedback mechanism or other relationship between development and capability among these states.

One might expect that, in the presence of a feedback effect, the best performers on military capability would lie near the upper end of the development scale, and the slopes of their regression lines would lie in the middle range, thus demonstrating some optimized ratio in converting GNP into military capability. In fact, the best performers on military capability are scattered throughout the range on GNP and GNP growth rate, and their slopes tend to be shallower than average. In contrast, the best performers on GNP tend to have the steepest slopes, to convert the least percentage of wealth into military power of all the group.

In digging beyond these average statistics, it is interesting to note what the GNP-military capability relationship does *not* vary with. For example, the slopes of the regression lines do not vary with the type of regime. Military regimes do not tend to have shallower slopes than average; they are no more given to concentrate on increasing military capability than civilian regimes. From the cases we examined, military rulers do not take power and then build up their armies, and military regimes do not necessarily lead to militarily powerful states.

The GNP-military capability relationship also does not vary with regime stability. The slopes are not correlated with either the number of coups or the number of coup attempts except to the extent that such instability is itself correlated with patron involvement. By the same token, the slopes do not correlate with other forms of domestic turmoil. Countries that have had horrible civil wars or episodes of ethnic violence are scattered randomly across the list of slopes. Nigeria's and Bangladesh's, for example, are steeper than average, Chad's and Uganda's are shallower than average, and Rwanda's and Burundi's are in the middle. This distribution indicates that domestic struggles for power, whether between individuals seeking to rule or between ethnic groups seeking power or autonomy, have not necessarily been an impetus to military-capability acquisition. There are no systematic differences between states like Malaysia, where the transfer of political power has not been problematic, and states like Uganda, where the fundamental structure of the regime has been at issue.

It also turns out that the relationship between GNP and military capability does not vary systematically with a state's location, size, or age or with its ethnic composition or the historical circumstances of its birth. Nor does the identity of the former metropole have any impact. Despite the very great differences in British and French colonial policy, despite the fact that the Belgians did so little and the Portuguese had to be thrown out by force, there is no systematic pattern in what has happened in their former colonies.

What the GNP-military capability relationship *does* vary with is economic performance, but it does so in a way opposite to that expected from the European model. The best performers do not lie around some optimized middle range of GNP-capability slopes, as the model would predict, but tend instead to cluster at the steep end. Those states with the largest GNPs, those that do best in their GNP growth rates, and especially those that made the greatest overall economic progress across the period are the states that paid proportionately the least attention to military capability. The fit is not perfect, to be sure, but it applies all across the GNP scale. The fact that the pattern holds across all levels of size and wealth suggests that there is no threshold above which relationships are different than they are below. It also suggests that we are not observing (at least so far) a group in transition, one in which the relationship between GNP and military capability changes as a state or regime matures.

Our examination of individual state histories showed that a feedback effect cannot be found even at a case-study level. Instead, we see a stair-step effect, at least in small states, where capability rises after several good years of GNP growth and then levels off after GNP stagnates or declines. There is no evidence of a causal mechanism between a need for military power and economic growth. On the contrary, many of the states that gain military power the fastest are the poorest performers economically.

To account for this pattern of development and military power in new states, we need to go back to the country-year studies, which dealt with all the data points in the set at one time. Recall the two central findings of that analysis: (1) that (controlling for patron effects) GNP and military capability are closely correlated at the country-year level, and (2) that patron support can permit a state to achieve high levels of military capability even in the total absence of economic growth. Together, these are essential clues to the individual state results.

In respect to the first finding, the European model would lead one to expect well-defined patterns of individual state behavior scattered all over the country-year chart; what we actually saw (Fig. 4.8) is relatively undefined patterns of individual state behavior on a very neat line up the diagonal of the chart. This violates a central assumption of the European model, because it is powerful evidence that no "law of the jungle" has operated for most new states over the past quarter-century. If it had, capability would vary with the security environment; states in hostile regions would have to acquire a lot of military capability to survive, and those in benign regions could afford to do little at all. Instead, utterly without regard to region, we find states arming about in proportion to their economic circumstances (again, controlling for patron effect). For most states, therefore, the external impetus to arm is absent, and although states do pursue military capability for any number of reasons, such as custom, ideology, prestige, or bureaucratic politics, they do so quite consistently as a function of their absolute wealth.

For some states, of course, a hostile environment or an aggressive foreign policy does provide the impetus to arm, and here the second major finding of the country-year analysis comes into play. There we saw clearly that the very poorest states can achieve levels of military capability as high as the very wealthiest with the support of a developed-state patron, especially if that patron is the Soviet Union.

Consequently, for states bent on increasing their military capability, there is an alternative to development—they can turn to a patron. This is most clear in the case of Somalia and South Yemen, two states that came into existence with expansionist foreign policies and built large forces with Soviet assistance. Their elites virtually ignored a development agenda, and their economies suffered accordingly, but military growth continued unabated. The finding also holds, however, in cases where the scale of assistance is less overwhelming. Both Kuwait and Malaysia are now rich enough to purchase needed military hardware and support substantial armies. But in their early years, each was saved from an external threat only by British intervention.

Finally, access to developed-state arms has an important impact even on wealthy states that purchase them at retail rather than receive them as gifts or at subsidized prices. States like Nigeria, Algeria, and Malaysia lack the industrial capacity to produce most modern weapons. Nevertheless, given their relative success at generating state income, they can buy virtually any weapon they desire. The availability of arms from foreign suppliers permits the transformation of national income directly into military capability without the need to develop the economic or political structures the European model assumes are required.

The result is that although states do, in fact, show variation in their GNP-military capability relationships based on the degree to which they are involved in external conflicts, the tendency is for those that are most conflict-prone to do less well on development indicators than average, not better. Their slopes are shallower, but they manage because of patron support. Those states well endowed with natural resources may grow quickly in both economic and military terms, but the military growth is able to keep pace only because of foreign sources of supply.

The relative power of developed states as suppliers of military hardware is hard to overstate. Recall that Somalia and South Yemen, two of the poorest states, had military forces as large as any in the group. Neither state, however, received more than a fraction of the total Soviet arms exports in any given year. The larger states of the Middle East (Syria and Libya, for example) all got far more. Similarly, the African beneficiaries of French assistance got only a very small percentage of total French arms exports, most of which went to Latin America, South Asia, and the Middle East. The United States generally makes smaller arms transfers to new states than the Soviet Union or Western Europe. The domestic market is so large that it can absorb most weapons production, so there is little incentive to "design down" to Third World cost and trained manpower limits. In addition, the United States produces armored vehicles, patrol boats, and tactical aircraft at a much slower rate than the Soviet Union, leaving less surplus for arms transfers to take up. Nevertheless, the United States, along with any one of a number of other developed states, continues to have the clear potential to quickly overwhelm local efforts to build up weapons stockpiles.

Summarizing: (1) the demands of international security have not been an impetus to development in modern new states; (2) the international environment for most has been benign, and for the others, patrons have provided for military assistance; and (3) at the individual state level, no feedback effect is observed; on the contrary, those states that pay the least attention to military capability do best in development. For most states, even military regimes, the armed forces seem an afterthought; for those to whom they really matter, alignment with a patron is most often the route chosen to gain military capability. The consequences for new states that result from these circumstances are often not very pleasant.

Consequences

If development in new states is not following the European model, what does that finding imply? One possibility, of course, is that the model has been inaccurate from the first, even for Europe. Certainly violent conflict between organized groups has been a feature of much of human history, but the modern state arose in only one epoch. To attribute that development to international competition, one must explain why such competition worked when it did, rather than a thousand years earlier or later.

The historians who specified the European model did not ignore this issue, although I have not addressed the point until now. Development in Europe proceeded as it did because such factors as the accumulation of scientific, technical, and geographic knowledge, growing populations, and rapid cultural change combined with the international political environment to produce a special set of circumstances. McNeill (1963), for one, argues that the result was a wholly new culture favoring individualism, a culture so unique that it will be as hard to comprehend a thousand years from now as it would have been a thousand years ago. Whether he is right or not, the fact that the model offers a plausible explanation for developing Europe does not mean it can only apply to the past; there are at least a few places where there are some clear signs the model's central thesis operates today.

The Republic of South Africa is an example of a country whose recent history seems to fit the European model quite well, and it is an especially interesting case because the decision to follow that model was taken in public. During the middle 1970's, sharp rises in the prices of gold and certain other minerals meant a windfall of foreign exchange earnings for South Africa, leading to a public debate about how the new wealth should be used. There were those in the ruling Afrikaner community who argued that the money should not be channeled into economic development because such development would of necessity result in more black skilled workers and managers and an increased black presence around white areas. This group preferred reduced economic circumstances to a higher standard of living if that higher standard brought with it changes in race relations. This was not a splinter position by any means, but the view of a major bloc of the National Party. It was a classic case of an elite that was content with its status and fearful of losing it if the economy expanded.

But in the end those who favored economic expansion prevailed, and one important argument they marshaled in support of their position is particularly revealing. This was the argument advanced by the defense establishment, which held that South Africa could no longer count on Western assistance (no patron) in the event of Soviet intervention (the United States had just refused to back up the Angola incursion) and therefore needed to beef up its economy in order to expand its defense industry. The victory of the pro-growth forces was quite complete; the defense minister, P. W. Botha, went on to become prime minister, the country underwent a period of rapid economic growth, and a successful program was carried out to increase weapons production capabilities in all areas. Although the case of South Africa is admittedly unique, not only because of its pariah status and its open political system (for the white minority), but because it had the luxury of an economic surplus to put to whatever use it chose, it clearly shows that the European model is not an irrelevant concept in the modern world.

If that is so, why should it not be relevant for modern new states? As noted back in Chapter 1, in size and population most new states are in the range of the proto-states of developing Europe, and they have the advantage of a far vaster fund of knowledge and technology available to them. Moreover, development, as an end in itself, is at least given lip service by nearly every new regime as a principal goal of the state. The great difference is the existence of developed states, a difference that is crucial and will not go away. The international system is so structured that it rules out the European model for the newcomers; the quantitative evidence is quite clear. The question is, how will this affect development? Will it forever stunt the process, have little impact, or drive it in new directions?

Answering the question begins with the recognition that there are at least some other routes to development besides the European experience. One can import the people and culture of a developed state and create a new developed state even in the absence of an external threat: New Zealand is one example; by some accounting, Canada might be another. The pressures to develop can also come from internal forces (e.g., the U.S. Civil War; various insurgencies around the world), although the record of success is spotty. It may be that an ideology of progress and development has grown so pervasive that regimes will be pressed willy-nilly into successful development without the external stimulus, but the findings of this research suggest caution in so hopeful a prediction.

The bottom line is that for the elites of new states, the presence of external patrons fundamentally alters the pressures to push development. In the European experience, elites had to make their countries work, not because of popular pressure to do so, but because if they failed they would be overthrown and conquered by someone who was better at nation-building than they were. Elites in modern new states have an alternative. For most, there is little in the way of an external threat; for the rest, patron support is a ready option for defense. Elites are still susceptible to the pressures of domestic politics, to be sure, but those pressures are not necessarily pressures to develop.

The consequences for the citizenry can be dreadful, not because people are deprived of a say in who governs (throughout most of history and in most places the common people have never had a say in such things), but because there are no pressures on those who govern to do a good job. Those pressures were not domestic in the past, and now they are not international either. Ethiopia stands as a tragic example. Famines have occurred in many places and times; sometimes governments do well in alleviating the hardships and sometimes they fail. When countries had famines before the modern era and handled them poorly, the governments were not voted out of office, but often neighboring countries carved their territories up for themselves.

Somalia tried such a move in the late 1970's, but it failed in the face of massive Soviet and Cuban military assistance to Ethiopia. Today Ethiopia is a disaster, but the regime is in no jeopardy. It has a huge, well-equipped army, which is maintained without any regard to the economic damages of the famine. The famine is, in fact, essentially irrelevant to the staying power of the regime, which depends on continued patron support and on the loyalty of the armed forces equipped by that patron.

A concern with maintaining the loyalty of the military is hardly unique to Ethiopia. The army's support is in fact a preoccupation of many elites, especially the support of the capital city garrison, and they must also always be concerned about keeping the capital's urban population peaceful, lest disorder open the way to military intervention. These can be difficult tasks, as the number of military coups among this group attests. The temptation is to cater to city consumers by keeping the price of food down, to the detriment of farmers, and to spend what national income is available in urban areas where the most important military units are normally stationed. Usually, neither policy is good for the country as a whole, but both can be perceived as important to the regime's survival. If there are international security problems, then the need to stay in a patron's favor also places pressures on the elite, although again not necessarily pressures to develop.

If there is an exception in this group, it may be Singapore, but because of its colonial history, it is not wholly comparable to the other states. Founded by the British, small, and with a largely Chinese population, Singapore was a major trading center long before the British left and so did not face the complications of development that other and larger countries faced. Still, Singapore and to a lesser extent Malaysia are the closest examples of new states in which the European model may be seen to operate. Their elites, under external threat, found the prospects of help from a patron limited, and a special urgency to development seems to have resulted. These Southeast Asian examples, like South Africa, suggest that the European model is relevant in modern times when the external conditions meet its assumptions.

Other than Singapore and Malaysia, those elites that have done the best economically (excluding those that benefited from windfall oil earnings) have been those that, contrary to the model, avoided creating a large military establishment and have not had external security problems. Here is evidence that, in the absence of an environment in which the European model can operate, the best growth takes place when defense is *not* an issue. Confidence that this observation is good news for new states must be tempered, however, because many of the new states in this category, despite their above-average performance, have still had weak and erratic growth records (especially in per capita terms) and unstable regimes. One can say that patron support in military capability does not aid development, but the absence of such support is no guarantee of development success.

The patron effect has decoupled international security from development. This certainly does not mean that all development efforts in new states are doomed to fail, but it does mean that one mechanism important in the past for enforcing competence on elites is missing for most new states. Other mechanisms are possible, of course, but whether they will be effective will depend in part on how the security environment for new states changes over the coming decades.

Forecasts

The country-level analyses showed what is happening, and the country-year-level analyses showed why; the system-level analyses indicated the trends. Recall that there was a remarkably linear growth in average military capability across the whole time span of the study (Fig. 4.5), and that the weapons stockpiles that generated this capability grew in equally linear fashion. Recall also that the growth in system inequality leveled off and essentially stopped by the early 1970's (Fig. 4.3). Military capability is increasing across the entire system and in essentially every country. The impact of all this military activity turns up in a variety of ways, and simple extrapolation of currently observable trends may not provide the best prediction of the future.

Nearly every modern new state gained independence with no capability for offensive military action and almost none for defense. Given the number of countries in the data set, there were remarkably few interstate conflicts compared with what might have been expected based on world history. But, this is not to say that the trend will continue. There is a threshold of military capability below which it is simply not feasible to wage war. It appears that in most states across most of the period, the threshold was not met; if that is so, it is a very good explanation for the phenomenon of so few conflicts in the face of so many disputes. With the linear increase in military capability taking place across the system, this state of affairs is unlikely to persist, and as more and more states cross the threshold of war capability, the relative peace observed to date may disappear.

There is some evidence to support the threshold theory even in the period studied. Of the ten new states with the greatest military capability, all but one have engaged in or been directly threatened by war with a neighbor. The exception is Nigeria, and it suffered through an enormous civil war. Hardly any such interstate conflict has been engaged in by weaker states, and essentially none by those in the bottom half of the group. Moreover, a look at the growth of military capability in individual states shows that none engaged in conflict until their military capability scores reached at least 1.0, but that the frequency of conflict increased dramatically, once a state passed this point. By 1981, more than a third of all new states had military capability scores exceeding 1.0, and projecting the average growth rate into the future suggests that by 1990 as many as two-thirds will exceed it. Military capability is a necessary, if not sufficient, condition for war, and we are likely to see this condition met far more frequently among new states in the future than in the recent past.

The question is what impact this increase in the capability for war will have. It can only make the rate of conflict go up. The data through 1981 quite clearly show that nearly every country with a high capability score had engaged in some form of armed conflict. It is true that this does not demonstrate causality. They may have armed in response to threats rather than gone to war because they were armed, but this is not relevant to the point at issue. There are more than enough unresolved disputes involving new states to provide reasons for many more conflicts. Whether the conflict begets the arms or the arms beget the conflict, they go together; the increase in military capability will most certainly be accompanied by an increase in conflict.

Will this violence, painful though it may be, at least spur development? It is unlikely to, and the culprit again is the patron, a phenomenon that not only has the effect of decoupling military power from development, but also has the paradoxical effect of both permitting conflict and at the same time preventing its resolution. For in cases of conflict, patronage, as we know, often provokes countervailing patronage from the opposite side in the East-West struggle; and though both sides are certainly interested in winning, both are also concerned that the conflict not escalate out of control, placing them at risk and costing them more than they wish to expend. Patrons very often give enough to permit the waging of war, but hardly ever enough to permit a permanent resolution of the dispute. Somalia still does not control all the land inhabited by Somalis, but its failure to win did not lead to its elimination, and its neighbors must still face its threats. A dispute is festering, but if war breaks out, the major powers, which have been responsible for keeping both sides armed, will never run the risk of permitting total defeat of one or the other.

As the future unfolds, new states may face the worst of all possible worlds in which they must endure the same violence and war as their European predecessors, yet without any development benefits to show for it. The data leave little room for optimism regarding an abatement in the growth of military capability or the spread of conflict. There is also little doubt that when security becomes an issue, for most states it will revolve largely around patronage, and that for many of the rest (those with oil revenues, for example), arms purchases rather than domestic production will be the route to military power.

Alternatives to this pessimistic forecast are possible. The developed world could radically limit arms transfers to new states, for example. The resulting international environment might be more benign than the environment that led Europe to develop, but this should hardly be viewed as a problem. The record of those states in benign regions that paid little attention to military capability suggests that development is certainly not impossible in such circumstances and usually proceeds better than in the patron-favored states.

It is true that restraint in arms transfers may not alter the fundamental developed-developing state relationship as long as security ties remain. For example, France's security guarantees to its former territories can have the same impact as security assistance would in the sense that these territories are under no pressure to develop their own resources to provide for military forces. On the other hand, to the extent that such ties limit arms transfers and thereby limit conflict, this may be the most promising of all the realistic alternatives.

It is also possible—and probably more likely than the option of developed-state restraint—that as development in new states goes forward, the costs of patronage will rise beyond what the developed world is willing to pay. Wars in the Middle East and Indochina have already demonstrated that at certain levels of size and political development, Third World states can engage in wars that are extremely costly in men and equipment, even by major power standards. Of course, the end of patronage is hardly the end of arms transfers. Recently, France passed the United States to move into second place behind the Soviet Union in the dollar value of arms sold to the Third World. Despite the French Socialist government's rhetoric to the contrary, economic pressures to export have driven the French arms industry to an increasing extent in recent years, and this pattern is likely to persist among the developed middle powers.

As long as the arms are paid for, the cost issue for the developed world disappears, and with it one major incentive for restraint. Among new states, the tie between security and wealth may be strengthened in such an event, if not the tie between security and industrial capabilities. Even so, this second alternative can hardly be considered a huge improvement on present trends. It may spur development, but the human costs of modern war are tremendous, and the impact on developing societies difficult to predict. Just the strains of absorbing modern military technology can threaten serious disruption to developing societies. We can put little confidence in cold calculations of the utility of large-scale modern war as an agent of development when there is so little experience among new states from which to predict; they may be utterly destroyed before they can ever master development.

Conclusions

The findings and forecasts that have flowed from this research can hardly be described as optimistic, and perhaps it is appropriate to conclude with some thoughts on the meaning of the work. We have looked at military power only on the state and global levels. But at the personal level its effects can be overwhelming. Consider the case of Sylvanus Olympio. Chapter 1 began with his words on the relationship between an army and independence, and ended by noting his demise at the hands of his own army. Even if the macro-level results of military interventions into politics are not very significant, the consequences for the individuals involved in systems where monopolies of the means of violence are maintained with such weak institutional constraints are serious in the extreme.

When military capability is applied to interstate war, the personal impact is multiplied ten thousand times. Yet even the effect of war can be eclipsed by the impact that a subsidized military buildup has on development in new states; and the data are quite clear that arms transfers inhibit rather than promote development. The human costs of failures in development can have far greater long-term effects than even large-scale non-nuclear war. The current famines in Africa are but one example of the horrendous consequences of botching the development process. It is likely that far more people will have died of starvation in the first 50 years of the postcolonial period than from combat in war.

None of this is meant to argue for a return to the continual warfare of early modern Europe and to suggest that new states cannot develop in the absence of some Darwinian competition between them. The terms of modern war strain such an argument past its breaking point. What has been offered here is, for the first time, *data* showing that the security circumstances of modern new states are different from those of their predecessors, and that these differences appear to have a marked impact on the development process.

It is important to consider the long-term consequences for development when there are few national security problems and those that do arise are managed through the manipulation of arms transfers by outside powers. Perhaps the Latin American experience may be a good, if disquieting, guide (limited mass participation, slower development, autocratic regimes). In the end, the question for developing countries is whether their patrons are really doing them any good when they provide military assistance and weapons trade opportunities. It can be argued that they are not, that military assistance produces dependence not independence, promotes stagnation not development, and results in insecurity not security.

Most of the new states in this data set were virtually unarmed when they came into being. This occurred not because of any high-minded idealism on the part of the metropolitan powers, but because they had settled their own disputes over colonial territories 60 years before and no longer had to worry about one another. The possibility that the new states of the world would constitute a system that excluded the instruments for the international use of force has disappeared, however, and the question of how such a system would have operated is moot. While the arming of these states was at first erratic, the concerns of their own elites and the intrusions of East-West competition have combined to ensure that the process has broadened and continued uninterrupted.

Arms control has been the subject of intense and passionate debate for over 40 years now, but the focus has almost always been on the major powers and nuclear weapons. In view of the data presented here, one might question this concentration and suggest that arms-transfer issues be given greater attention. Nuclear weapons are certainly important security concerns and, indeed, can be a danger to all human life, but they have not been used in war since 1945, and they cost only a small fraction of the global wealth. In contrast, the Third World has seen extensive and continuing warfare in which literally millions have died, the majority of them killed by means of First and Second World weapons and technology, in which uncountable wealth has been expended, and in which the development process has demonstrably not been helped.

East-West competition is now considered too dangerous to conduct very near to the parties directly involved, but the fallout for the rest of the world has been a flood of weapons loosed to balance trade or win points in a competition wholly irrelevant to the people most affected. The result has been a process of arms transfers that may be the most counterproductive activity in the whole spectrum of relations between the developed world and new states.

The irony is that the arms suppliers hardly understand or give a thought to the long-term impact of their actions. The harm done is (one suspects) not deliberate, and the fact that the regions involved are so rarely essential to the central security interests of developed states makes it especially pointless. Perhaps this offers a hope that change is possible, that some mutual restraint by all suppliers can limit arms transfers to the developing world. It certainly will not bring peace, and it is unlikely to recouple the struggle for security to the drive for development; but it may limit the intensity of conflict to what developing societies can endure, and by doing so it may offer a breathing space for them to succeed in the modern world. It also may be the only exit from the present trap that leaves new states open to the devastation of war at the same time it inhibits the fundamental foundations of the development process.

Reference Material

APPENDIX A

Weapons Categorization

Data were collected on the annual totals of 325 different weapons systems in the military forces of 46 new states for the years 1957-81. The first step in constructing the military-capability index was to reduce this large number of separate weapons systems into a smaller number of groups based on mission and capability. Thirteen weapons categories were defined so as to include all weapons on which data were collected and to group them by mission in such a way that each state could reasonably be expected to have an interest in filling every category. The exceptions, of course, were the naval categories for the cases of landlocked states, but these were dealt with in a special way.

Most weapons can be employed for a variety of missions. In assigning a category, the coding scheme used here first considers the primary role for which a system was designed and then gives attention to its potential in other roles. For example, attack aircraft are defined as planes used for air-to-ground ordnance delivery and fighter aircraft as planes used for air-to-air. Air-to-air delivery is a more technically complex and aerodynamically demanding mission, so that fighters are generally more expensive and higher capability weapons than attack aircraft. They generally are assigned a secondary air-to-ground role, and sometimes models are produced explicitly for the ground attack mission, but they represent a different category of weapon. In essence, therefore, the categorization used captures the overall capability of the weapons system as well as its nominal assignment. Since the aircraft coded as fighters are more expensive than attack aircraft, even wealthy states would be well advised to invest in some of the cheaper aircraft for the air-to-ground mission and so keep both categories filled. It turns out, in fact, that this is generally the case.

Listed below are the 13 weapons categories, the rules used for assigning weapons to them, and a list of the weapons systems assigned to each category. The originating country is also listed. Where two or more countries are given, they are either licensed producers or retransferers of previously acquired systems. The data on military personnel and weapons stocks collected for this research are available through the Inter-University Consortium for Political and Social Research.

1. Utility Aircraft

Definition. Light fixed-wing and helicopter aircraft used in light transport, vertical assault, artillery spotting, and other support roles.

Coding rules. Payload: less than 15 passengers or 5,000 pounds of cargo. Range: less than 500 miles. Performance: subsonic. Ordnance: 1,500 pounds or less external carriage.

Models: De Havilland DHC-2, DHC-3, DHC-4, DHC-6 (Canada); Aerospatiale S.A. 318, S.A. 341 Gazelle, S.A. 360/365 Dauphin (France); Corvette (France); Dassault M.D. 312/315 (France); Max Holste M.H. 1521 (France); MB 105 (France); Rallye Guerre (France); Aerfer-Aermacchi AM3.C (Italy); Aermacchi A.L. 60 (Italy); Piaggo 149D (Italy); De Havilland Devon (New Zealand); Turbo Porter (Switzerland); Austor AOP mk 9 (UK); Britten-Norman BN2-A (UK, Romania); De Havilland DH-104 (UK); Hawker Siddeley H.S. 125 (UK); Hunting CMK.1 (UK); Scot Twin Pioneer (UK); Short Skyvan 3M (UK); Westland Whirlwind (UK); Aero Commander (USA); Beechcraft 18/C45 (USA); Beechcraft Queenair/Kingair (USA); Bell 47G, 204, 206 (USA); Cessna 185/402, 310/U3 (USA); Fairchild Hiller FH-1100 (USA); Piper Aztec, Piper Cub, Piper Navajo (USA); Sikorsky S-55 (USA, France); Sikorsky S-58 (USA, France); Antonov AN 2 (USSR); MIL MI-4 (USSR); Yakovlev YAK 12A (USSR, Poland); Dornier Do-27, Do-28 (West Germany).

2. Transport Aircraft

Definition. Medium and heavy-lift aircraft capable of hauling troops and cargo over long distances; may be fixed wing or helicopter.

Coding rules. Payload: more than 15 passengers or 5,000 pounds of cargo. Range: more than 500 miles. Performance: subsonic; may be jet or prop driven. Ordnance: not applicable.

Models. CL-215 (Canada); De Havilland DHC-5 (Canada); Aerospatiale S.A. 316, S.A. 321 (France); Aerospatiale/Westland S.A. 330 (France, UK); Caravelle (France); Dassault Falcon 20 (France); Nord .2501 (France); Nord N262D Fregate (France); Transall C.160 (France, West Germany); NAMC YS-11A (Japan); Fokker-VFW F.27, Fokker F.28 (Netherlands); De Havilland DH114 (UK); Hawker Siddeley 748 (UK); Hunting Herald (UK); Westland mk53, S-61 (UK); Boeing CH47, 727, 737 (USA); Curtiss-Wright C-46 (USA); Douglas C-47 (USA, France, Australia, Iran); Douglas C-54, C-118, DC9 (USA); Grumman Gulfstream (USA); Lockheed C-13O (USA); Antonov AN 12/26, AN-24 (USSR); Ilyushin Il-14M, Il-18 (USSR); MIL MI-4 (USSR); Yakovlev YAK 40 (USSR).

3. Attack Aircraft

Definition. Low-to-medium-capability aircraft usually employed to deliver air-to-ground ordnance with only limited air-to-air capability.

Coding rules. Payload: more than 1,500 pounds of armaments. Range:

more than 150 miles. Performance: less than 500 knots in level flight. Ordnance: capable of both forward firing and gravity munitions; no all-weather delivery capability.

Models. GAF 22 Nomad (Australia); EMB-111 (Brazil); Reims-Cessna 337 (France); British Aircraft BAC 167 (UK); Britten-Norman Defender (UK); Douglas A1D (USA, France); Martin B26 (USA); North American F51 (USA); OV-10 (USA); MIL MI-24 (USSR); Soko J1 (Yugoslavia).

4. Fighter Aircraft

Definition. High-capability aircraft generally designed for air-to-air combat but almost always with air-to-ground capability as well.

Coding rules. Payload: more than 4,000 pounds of armaments. Range: more than 250 miles. Performance: more than 500 knots in level flight; must have all-weather flying capability but not necessarily all-weather weapons delivery capability. Ordnance: capable of carrying both air-to-air and air-to-ground weapons, the latter including both forward firing and gravity munitions.

Models. North American F86 (Australia); Dassault Mirage III, V, F1 (France); Marut HF-24 (India); Fiat G91 (Italy); British Aircraft BAC Lightning (UK); De Havilland DH 113 Vampire (UK); Hawker Siddeley Hunter (UK); Douglas A4 (USA); Northrop F5E (USA); Ilyushin Il-28 (USSR); Mikoyan-Gurevich MiG-15, MiG-21, MiG-23, MiG-25 (USSR); MiG-17 (USSR, PRC); MiG-19 (USSR, PRC); Sukhoi SU-7, SU 20/22 (USSR).

5. Trainer Aircraft

Definition. Aircraft used principally for instruction but generally capable of limited light attack roles.

Coding rules. Payload: at least two pilots, dual controls with instructional layout. Range, performance, ordnance: not applicable.

Models. Canadair CL-41A (Canada); De Havilland DHC-1 (Canada); Aero L29 (Czechoslovakia); Ziln (Czechoslovakia); Dassault-Breguet/Dornier Alpha Jet (France, West Germany); Potez-Air Fouga C.M. 170 (France, Israel); HAL HT.2 (India); HAL Pushpak I (India); Aermacchi MB.326 (Italy); Piaggio P.148 (Italy); SIAI-Marchetti SF.260 (Italy); AESL Airtourer (New Zealand); BT6 (PRC); SAAB MIF-15 (Sweden); British Aircraft BAC 145.T52 (UK); British Aircraft Lightning T54/T55 (UK); De Havilland DHT-1 (UK); Hawker Siddeley Hunter T.77 (UK); Hunting T.mk 1 (UK); Scot Bulldog (UK); Beech T34/F33 (USA); Cessna T41/172/180/182 (USA); Douglas TA4 (USA); Hughes 269A/300 (USA); Lockheed T33 (USA); North American T6 (USA, Belgium); North American T28 (USA); Mikoyan-Gurevich MiG-15 UTI, MiG-21 UTI (USSR); Yakovlev YAK 11, YAK 18 (USSR); Soko G2A (Yugoslavia).

6. Armored Cars

Definition. Protected transport and attack vehicles for use in counterinsurgency and infantry support missions.

Coding rules. Capacity: 7 or fewer soldiers. Armament: 70mm gun or smaller. Drive: Wheeled or half-tracked.

Models. Armored car (Egypt); AML 60 (France, Belgium); AML 90 (France, Belgium); Cascavel (France); EBR 75 (France); M3A (France); Mowag (France); Panhard (France); Dailmer (UK); Ferret (UK, South Africa); Fox (UK); Saladin (UK); Saracen (UK); Shoreland mk52 (UK); UR416 (UK); V 100, V 150, V 200 (UK); Vicker mk3 (UK); M-8 Greyhound (USA, France).

7. Armored Personnel Carriers

Definition. Armored troop transport vehicles capable of carrying more personnel than armored cars and with generally greater protection.

Coding rules. Capacity: 8 or more soldiers. Armament: 70mm gun or smaller. Drive: Wheeled or tracked.

Models. OT-64 (Czechoslovakia); AM-VIII, Savien, VXB-170 (France); K63 (UK); M113 (USA); BMP, BRDM, BRT-40, BRT-50, BTR-152 (USSR).

8. Tanks

Definition. Armored vehicles carrying a large gun that can be aimed without altering the heading of the vehicle.

Coding rules. Capacity: 3 or 4 crewmen. Armament: at least a 71mm gun. Drive: Tracked.

Models. AMX 13 (France, Israel); Sherman (Israel); T-59 (PRC); Centurion, Chieftain, Comet, M24, Scorpion, Vickers 37T, Vickers mk3 (UK); PT-76 (USSR); T-34 (USSR, Czechoslovakia); T-54 (USSR); T-62 (USSR, PRC).

9. Missiles

Definition. Ground-launched guided missiles with either anti-air or anti-armor missions.

Coding rules. Guidance: must be guided after launch either by ground-based system or by self-contained homing equipment. Launcher: may be launched from either fixed site or by individual soldiers, but must be ground-launched (i.e., not exclusively air- or sea-launched).

Models. Savin SS-11 (France); British Aircraft BAC Vigilant, Rapier (UK); Raytheon Hawk (USA); AT3 Sagger, Frog-4, SA-7 (USSR); SA-2 (USSR, Egypt).

10. Guard Boats

Definition. Small craft having very light weapons and only an inshore patrol or minesweeping capability.

Coding rules. Size: under 50 tons standard. Speed: under 20 knots. Armament: 23mm guns or smaller.

Models or classes. PC (East Germany); motor launch (Egypt); coast guard PC, LCT, motor launch, new launch (1938), river patrol boat, 20-ton PC, 70-ton river patrol boat (France); LCVP (Gabon); old PC (Germany); LCM (Ivory Coast); launch, river PC (PRC); police launches (2 classes; Singapore); coastal PC, Ford, Ham, HDML, motor launch, repair launch, river patrol boat, supply boat, survey vessel, tender, Ton, yacht, 39-ton PC (UK); auxillary LST, coastal minesweeper, motor launch, river patrol boat, Swiftboat, tender (USA); AGI, P4, Poluchat I, T58 minesweeper (USSR); auxillary tender, patrol boat (West Germany).

11. Patrol Craft

Definition. Vessels capable of limited offshore patrol roles, possessing either very high speed or at least medium size for some seagoing missions and weapons with some anti-ship capability.

Coding rules. Size: more than 50 tons or more than 35 tons if speed greater than 20 knots. Speed: more than 14 knots; must be more than 20 knots unless size is greater than 50 tons. Armament: at least 23mm guns, or guided missiles or torpedoes.

Models or classes. Acute (Australia); Franco-Belgian 235T, VC, 40-ton PC, 128-ton PC, 150-ton missile boat, 180-ton PC, 280-ton/40mm gunboat, 376-ton/76mm gunboat, 385-ton Exocet (France); 85-ton PC (Gabon); large PC (Netherlands); Huchwan, Shanghai I, Shanghai II, Shershen (PRC); 40-ton missile boat, 240-ton missile boat (Singapore); 62-ton PC, 96-ton PC, 96-ton missile PC, 100-ton PC, 100-ton/76mm PC, 115-ton PC, 120-ton missile boat, 130-ton missile boat, 139-ton/40mm gunboat, 140-ton PB, 155-ton gunboat, 160-ton/40mm gunboat, 200-ton PB, 354-ton gunboat (UK); Subchaser (USA, France); Komar missile boat, MO-VI, Osa missile boat, P6, SO-I, Zhuk PB, 255-ton PB (USSR); 190-ton PC (Yugoslavia).

12. Corvettes

Definition. Ocean-going vessels with major anti-ship, anti-submarine, or anti-air capabilities.

Coding rules. Size: more than 350 tons. Speed: more than 15 knots. Armament: at least 40mm guns, plus either surface-to-surface missiles or anti-submarine weapons (depth charges, missiles, or torpedoes).

Classes. 850-ton corvette, 1724-ton frigate (Netherlands); 375-ton corvette, 440-ton corvette, 500-ton corvette, 770-ton corvette, 1250-ton frigate, 1700-ton frigate, 1800-ton frigate, 2300-ton frigate, 2400-ton frigate, 3630-ton frigate (UK).

13. Amphibious Craft

Definition. Vessels designed to transport combat troops and supplies to shore without the need of docks or lighters.

Coding rules. Size and speed: not considered; must be able to carry combat troops and supplies to beach.

Models or classes. LCT (France); LCVP (Gabon); LCM (Ivory Coast); assault craft (PRC); LCM, LCP, LCT, 810-ton LCT (UK); LST (USA); LCU, Polnoeny LST, Ropucha LST, T4 LCVP (USSR).

APPENDIX B

Conjoint Measurement

The analysis required an algorithm capable of transforming annual national data on military personnel and 13 categories of weapons systems into a single annual national military capability score, and of doing so in accordance with the index construction rules specified for the study: the more the better, and the more balanced the better. Those rules eliminate simply summing across all categories and require that the algorithm accommodate data in different units (people, tanks, etc.) and spanning different ranges (3,000 to 763,000 for people; 0 to 725 for tanks, etc.). Among a variety of nonmetric programs the one most suitable for this study is Conjoint Measurement III (CM-III) from *The Guttman-Lingoes Nonmetric Program Series* (Ann Arbor, Mich., 1973). CM-III was employed twice in this analysis: first separately on naval and non-naval categories, and then again on the results to compute a final index.

Although the categories to be scaled must be rank ordered for CM-III, they do not have to have any known linear relation to each other or to the underlying capability dimension one wishes to measure. Both conditions were met in this case. The algorithm is designed to transform the variables so that they are linearly related both to each other and to the capability dimension, and to calculate a single "best score" that taps that capability dimension. It does this by differentially shrinking and stretching the intervals between observations until an optimal scaling is achieved (i.e., when linearity between components is maximized), subject to the constraint that if a variable input to CM-III has a higher score for case A than for case B, then the final transformation of the variable will also have a higher value for case A than for case B. The transformed scores will thus be monotonic with the original data, that is, rank ordering of the cases on each dimension will remain unchanged, and the combined score will be linearly related to each.

The best way to illustrate the operation of this algorithm is with a simple example (Stoll 1977). Table B.1 shows the number of fighter planes, military personnel, and armored cars possessed by states A-E. Since the variables are in different units across different ranges, in the first step the variables are stan-

TABLE B.1
Hypothetical Military Capability for Five States, Raw Data and Z Scores

	Fighters		Military personnel		Armored cars	
State	Number	Z score	Number	Z score	Number	Z score
A	1	−.768	2,676	−.080	433	−.751
B	40	−.508	1,774	−1.540	455	−.735
C	76	−.267	3,050	.339	563	−.652
D	377	1.740	2,859	.058	2,432	.766
E	86	−.200	3,651	1.220	3,230	1.370

dardized (Table B.1, z scores). The algorithm then calculates the best predictor of each state's z score—which, using least-squares criteria, is simply the mean of its score across all variables. CM-III then standardizes these means across each state yielding the following results:

Variable	State A	State B	State C	State D	State E
Mean of original z score	−.533	−.928	−.194	−.856	.799
Z score of mean	−.667	−1.160	−2.420	1.070	1.000

The second row of this tabulation represents the best predictor of the underlying capability dimension, assuming that the relationship between capability and the three z-scored variables was linear, that each variable was weighted equally, and that the composite score is an additive combination of the component variables.

Given these best predictors, the algorithm needs a method to stretch and shrink the intervals on each variable so that they will be more linear with each other. CM-III uses Guttman's rank-image principle to perform this stretching and shrinking. It operates by taking a variable (fighters), finding the highest score (1.74 for state D), and replacing this highest score with the highest score from among the z scores of the means (1.07). The process is repeated for the next-highest score and so on for this and all other variables until each variable score has been replaced by the z score of the means with the corresponding rank.

Using the transformed scores this time, the algorithm again calculates a mean score for each state and standardizes these means to yield a better estimate of the capability score than before. Table B.2 displays the transformed scores, their means, and their standardized means.

The average intercorrelation using the raw data was .457 between indicators, and after one iteration of the CM-III algorithm it rises to .690. The operation continues with further iterations identical to the first until a peak is reached in the intercorrelation scores or 100 iterations take place, whichever comes first. Were this example an actual case, a perfect fit would eventually be obtained with an average intercorrelation of 1.00. Table B.3, which charts the progress of the algorithm, helps to show how the results relate to the original data.

What CM-III has done is collapse the intervals between A and B at the low end and C, D, and E at the upper end. A and B were consistently lower than the other states on all scores and so they are ranked that way in the end. C, D, and E were not so consistent, and their scores turn out to be quite close together. They gain little advantage from being very strong on any single dimension without corresponding strength on the others.

Using the algorithm for this data set required modifying the original Guttman-Lingoes program to accommodate a larger matrix than that for which it was originally written (500 cases maximum). The original program did not assume that all scores must move in the same direction as the underlying dimension; that is, if manpower went consistently down as weapons stocks went up, it would assume that capability increased as manpower decreased and reflect that variable (so that less would mean better for manpower). This would violate one of the specified index construction rules on which these calculations were done (more is better) and required further modification of the program to disable the subroutine that causes reflection. It turned out that no category contained a net average negative correlation, however, and so no reflections would have occurred in any case.

CM-III has been used before to generate indexes of national capability (Gochman 1975), but only with respect to such broad indicators as population, GNP, and steel production. Its use here required careful attention to categorization, because if all states could not be expected to have some interest in each category, then they would be unfairly penalized by the algorithm for an unbalanced force structure.

The categories were arranged with this restriction in mind. In order to compensate for the lack of naval vessels among the 12 landlocked states, the algorithm was run in three parts. First, a composite score was calculated for each of the 859 country-years in the study using military personnel and nonnaval weapons categories only. Then the process was repeated for naval weapons for just those 621 country-years involving the 34 states with seacoasts.

TABLE B.2

Transformed Scores, Means, and Z Scores of Means for Five States

State	Fighters	Military personnel	Armored cars	Mean z score	Z score of mean
A	−1.160	−.667	−1.160	−.996	−1.120
B	.667	−1.160	.667	−.831	−.933
C	−2.420	1.000	−2.420	.172	.193
D	1.070	−2.420	1.000	.609	.683
E	1.000	1.070	1.070	1.050	1.180

TABLE B.3
Successive Estimates of Composite Capability for Five States

State	Input data	After one iteration	Final estimates
A	−.667	−1.120	−1.1250
B	−1.160	−.933	−1.2250
C	−2.420	.193	.8161
D	1.070	.683	.8165
E	1.000	1.180	.8169

The task was then to generate artificial navy scores for the 12 landlocked states proportional to their non-naval scores. Least-squares regression produces the maximum likelihood estimator of the relationship between two variables such as these, and so regressing the non-naval and navy scores yielded the best formula for calculating the missing navy scores from the corresponding non-naval ones.

Inserting the artificial scores for the 238 landlocked country-years then resulted in a complete list of 859 naval scores. The algorithm was then rerun using the non-naval score for each of the ten non-naval variables and using the naval score for the three naval variables, thus preserving the original weight of each category. The effect of this procedure was that the landlocked states were ranked only across non-naval categories, while the others, on the same index, were ranked across all categories.

The CM-III algorithm provided the statistical tool needed to transform the data on personnel and weapons stocks into a single composite index based on the quantity and balance of each state's military force. By calculating the index for the entire 859 country-year sample as a block, rank order was preserved across the entire data set, so that a tie score results only if the raw figures do not change between years. In this way, the final index provides a measure of each state's growth in military capability and not just an indication of its annual relative position. If the index were calculated in annual increments, the results would not have provided a uniform scale for judging the growth rate of individual states and would have displayed only annual changes in a state's relative ranking.

Bibliographical Note

The data on military personnel and weapons stocks were assembled from a variety of sources falling into three broad categories: publications that provided technical data on weapons systems, publications that provided annual arms-transfer figures, and publications that provided actual personnel and weapons stocks totals by country, though generally not on an annual basis.

Because annual weapons stock data were not available for all countries for the entire period, annual arms-transfer totals were integrated over time to produce the stock totals, and these were then updated with actual weapons-stock figures where possible. This procedure worked quite well. Updating required only minor alterations to account for aircraft and vehicles lost in accidents.

The principal sources used are set out below, together with some comments on their strengths and weaknesses.

Nicole Ball, *Third World Security Expenditure: A Statistical Compendium* (Stockholm, 1984), which summarizes military expenditure and personnel data from the late 1950's through the early 1980's for 48 Third World states, is especially useful because the author fully documents her coding rules and explains her reasons for choosing them. Although one might quibble with some of the coding decisions, one is never left to guess at how the numbers were arrived at. In this respect, the work stands alone. The problem is that the expenditure data are in local currency, making their incorporation into a cross-national data base of limited utility. Nevertheless, the discussion of expenditure data collection and coding is so well done that for any other purpose this volume is the best in its class, subject to the limitation that only 48 states are covered.

Trevor N. Dupuy, Grace P. Hayes, and John A. C. Andrews, *The Almanac of World Military Power,* 4th ed. (Novato, Calif., 1980), along with the earlier editions of 1970, 1972, and 1978, provides the best all-around summary of military manpower and equipment in Third World states. The authors not only give numerical totals, but also give breakdowns on the command structure, figures on the foreign forces in the countries, and the size of the national police forces. In addition, a brief history of the internal and external security situation is provided for each state, along with some general population and economic development figures. The only shortcomings of these volumes are that the earliest edition dates back only to 1970, and they have not been issued annually. Although most data seem to be current as of the end of the year before publication, the authors do not indicate where this is not the case.

Laurence L. Ewing and Robert C. Sellers, eds., *The Reference Handbook of the Armed Forces of the World* (Washington, D.C., 1966), is the only comprehensive source on Third World military forces compiled before 1970. As such, it is an important check on the figures calculated by adding up arms-transfer totals and also very helpful for comparing the military organization in many new states shortly after independence with that of later years. Later editions were published in 1971 and 1977, but by then more comprehensive coverage was available from other sources. The definition of military personnel appears to be narrower than in the publications of the U.S. Arms Control and Disarmament Agency (see below), because the ACDA usually shows higher figures. But this might in part reflect weaker intelligence during the early and middle 1960's, since in subsequent editions most sources, including the ACDA, have raised many more manpower estimates for this period than they have lowered.

David Harvey, ed., *Defense and Foreign Affairs Handbook* (Washington, D.C., 1980), is a comprehensive reference that presents the order of battle for every nation in the world, plus lists of government leaders and selected economic and government statistics. It seems to use the same sources as *The Military Balance* (see next entry), but its coverage is more detailed and is geographically broader. Although it gives the details of military organization, it is weak on stock totals, generally listing only the type of weapon with which individual units are armed. It is quite valuable for the year of its issue (there is an earlier, 1976 edition), but it is not an annual publication.

The International Institute for Strategic Studies, *The Military Balance* (London), issued annually since 1960, provides a breakdown of the size, budget, force structure, major weapons stocks, and deployment of forces for about half the states of the world and a more limited store of information on many of the rest. It contains some of the most thorough and complete national data available, but its coverage, especially for the early years, is limited. Nonindustrial states have been covered in significant numbers only since 1971, so that it is not useful for Third World studies in earlier years. Since this annual seems to be most accurate about a year after any major changes have taken place, it is very helpful as a check on other sources. However, it is of limited use in time-series research because it does not give revised estimates for previous years.

John Keegan, ed., *World Armies* (London, 1979), is an exceptional work that contains a comprehensive history of every national army whose order of battle it catalogues—152 countries all told. Although the weapons stock data are less comprehensive than in other sources, the work is unmatched in its discussion of deployments and organization. Keegan assembled a team of 16 experts (most from the Royal Military Academy, Sandhurst) for these country studies, and though the British bias in the military histories is clear, he has produced an invaluable resource.

Nikolas Krivinyi, ed., *World Military Aviation* (New York, 1977), provides a detailed and accurate description of the organization, bases, aircraft, procurement plans, and military assistance relationships of virtually every national air force in the world—128 in all. Equally important, it contains a half-page account of the size, performance, and production history of every aircraft type and major variant found in those 128 air forces. It is the single best reference for aircraft performance because of its brief and uniform format. It also provides extremely detailed descriptions of national air weapons stocks, but only for the year (presumably) before publication and with no historical material.

Jean Labayle Couhat, *Combat Fleets of the World* (Annapolis, Md.), is an English translation of a French work (*Flottes de combat*) that is published biennially. The French volumes date back to the end of the Second World War, but the English series did not begin until 1976. The work is similar in format to *Jane's Fighting Ships* (see below), but is less exhaustive in its description of vessels. It is an invaluable source, however, for those former French dependencies that have acquired navies. It is more accurate than *Jane's* for those states in their first two decades, especially on delivery dates and suppliers. It makes a very good complement to the larger volume, with which it agrees in every detail except for the Francophone world.

J. E. Moore, ed., *Jane's Fighting Ships* (London), is a yearbook that needs no introduction. It is universally recognized as the best source on naval ships worldwide. It is especially easy to use because it is organized by country rather than by ship type. It contains a very detailed description of virtually every warship in service. Its weaknesses are minimal but ought to be mentioned. Although the technical data are consistently good, additions and deletions are most accurately recorded two years or so after they occur. This is probably due more to changes in the schedules of various countries than sloppiness on the part of the editors, but the accuracy improves only as the time of interest recedes. Of more concern, the data on the smaller Third World navies are incomplete and sometimes inaccurate in the pre-1970 yearbooks, especially for those states that were not former British possessions. For those states, some additional reference material (e.g., *Flottes de combat;* see Labayle Couhat, above), is usually required to pin down the exact date of early vessel acquisitions.

Two volumes that are excellent sources for technical information on all major infantry weapons produced or used by NATO and Warsaw Pact countries are John I. H. Owen, ed., *NATO Infantry and Its Weapons* (Boulder, Colo., 1976), and John I. H. Owen, ed., *Warsaw Pact Infantry and Its Weapons* (London, 1976). I did not encounter one infantry weapon in this research that was not covered in these volumes. They provide little information on the transfer of weapons as such but are invaluable for coding those mentioned in the transfer-data sources.

Ruth Leger Sivard, *World Military and Social Expenditures* (Leesburg, Va.), is a comprehensive yearbook (issued since 1974) that largely covers the same ground as the ACDA annuals (see below) but pays more attention to repairing errors in the data of past years. This makes it especially valuable for time-series research. Its most recent figures, like those in most sources, are not as accurate as those a few years old. It contains no data on hardware, but provides a great variety of expenditure and personnel figures. Although the sources for the numbers are not always clear and the political biases of the author come through strongly in the commentary, it seems to be at least as accurate as the ACDA publications, and since these were discontinued after 1982, Sivard's annual is likely to become the standard reference.

The Stockholm International Peace Research Institute's *SIPRI Yearbook of World Armaments and Disarmament,* published annually since 1969, contains a so-called arms trade register that lists the recipient, supplier, number and type of weapon, order date, and delivery date for all transfers of major weapons known to have taken place during the preceding two or three years. There are also several separately published editions of these registers for cumulative transfers going back to the mid-1950's. These SIPRI publications were principal sources for this research. Totals gleaned from them required little revision when matched against available single-year stock totals through 1975. Their data, however, are not fully reliable for transactions occurring within two years of publication, since deliveries are often not made or orders filled as originally proposed. Since 1975, moreover, the data have been less comprehensive on Third World states than in earlier volumes, either because SIPRI has fewer resources to devote to collecting data from those areas or because it considers the matter less important now than in the first years of independence. One other significant weakness is that there are no records of the transfer of field guns, even in the very largest calibers. Other sections of the yearbooks deal with arms aid, but the cost figures, as discussed in Chapter 3, are of limited value.

J. W. Taylor, ed., *Jane's All the World's Aircraft* (London), is another yearbook full of detailed information, providing technical data on all aircraft in production (both civil and military) anywhere in the world. But these volumes are of little use in determining weapons stocks or transfer rates, because aircraft are listed by country of production and manufacturer rather than by country of ownership. Every aircraft is not listed every year, but each volume has an index listing all types and the edition and page number where they can be found. *Jane's All the World's Aircraft* is most useful for filling in the gaps left by the Krivinyi work (which fails to cover every aircraft possible, since some had left service before 1977), but it is quite cumbersome as a source in cross-national studies.

Finally, as noted above, the United States Arms Control and Disarmament Agency's publication, *World Military Expenditures and Arms Transfers* (Washington, D.C.), has ceased publication. Available in annual editions from 1971 through 1982, these volumes provide data by country and region on GNP, military expenditures, and a variety of social expenditure data for ten-year periods. Transfer sums are listed but only by region. The publications also provide a most useful compendium of population and military personnel data by both country and region. Although these works have the most complete country coverage on a broad range of variables, the data were of limited utility for several reasons. For one thing, though the figures go back as far as 1961 (in the 1971 edition), errors that crop up in previous yearly totals are often not corrected. Consequently, the current edition may have the most accurate figures, but there are often major discontinuities from one year to the next. Another problem has to do with the military personnel figures. Although I used these volumes as my principal source on troop strength, many of the data had to be amended from less-comprehensive sources because of a lack of rigor in the ACDA coding rules. Those rules are specified only generally and exclude various national police forces without specifically mentioning where they are excluded. In order to employ these data with confidence it is necessary to find a more detailed source for at least one year and decipher the coding decisions of the U.S. Defense Intelligence Agency (the source of ACDA's figures) by comparing the breakdown of figures in that source with the ACDA total. If this is done, then *World Military Expenditures and Arms Transfers* is a most useful reference for filling in personnel data on the many countries and years sparsely covered by other sources.

Bibliography

Almond, Gabriel, and James S. Coleman, eds. 1960. *The Politics of Developing Areas.* Princeton, N.J.: Princeton University Press.

Arlinghaus, Bruce E. 1984. *Military Development in Africa.* Boulder, Col.: Westview.

———, ed. 1983. *Arms for Africa.* Lexington, Mass.: D. C. Heath.

Attwood, William. 1967. *The Reds and the Blacks.* New York: Harper & Row.

Ball, Nicole. 1984. *Third World Security Expenditure: A Statistical Compendium.* Stockholm: National Defence Research Institute.

———. 1985. "Defense Expenditures and Economic Growth," *Armed Forces and Society* 11.2 (Winter): 291.

Bebler, Anton. 1973. *Military Rule in Africa: Dahomey, Ghana, Sierra Leone, and Mali.* New York: Praeger.

———, ed. 1981. *International Political Science Review* 2.3.

Bell, J. Bowyer. 1973. *South Arabia: Violence and Revolt.* London: Institute for the Study of Conflict.

Benedict, Burton. 1965. *Mauritius: Problems of a Political Society.* London: Praeger.

Bennett, George. 1964. *Kenya, a Political History: The Colonial Period.* London: Oxford University Press.

Benoit, Emile. 1973. *Defense and Economic Growth in Developing Countries.* Lexington, Mass.: D. C. Heath.

Biennen, Henry. 1974. *The Politics of Participation and Control.* Princeton, N.J.: Princeton University Press.

———, ed. 1968. *The Military Intervenes: Case Studies in Political Development.* New York: Sage.

Biskup, Peter, B. Jinks, and H. Nelson. 1968. *A Short History of New Guinea.* Sydney: Angus & Robertson.

Bunge, Frederica M., et al. 1980. *Cyprus, a Country Study,* 3d ed. Washington, D.C.: American University.

Butterworth, Robert Lyle, and M. E. Scranton. 1976. *Managing Interstate Conflict, 1945-1974: Data with Synopses.* Pittsburgh: University Center for International Studies.

Callaghy, Thomas M. 1984. *The State-Society Struggle: Zaire in a Comparative Perspective.* New York: Columbia University Press.

Carter, Gwendolen M., ed. 1963. *Five African States: Responses to Diversity.* Ithaca, N.Y.: Cornell University Press.
Castagno, Margaret. 1975. *Historical Dictionary of Somalia.* Metuchen, N.J.: Scarecrow Press.
Clapham, Christopher S. 1976. *Liberia and Sierra Leone: An Essay in Comparative Politics.* New York: Cambridge University Press.
Cox, Thomas S. 1976. *Civil Military Relations in Sierra Leone.* Cambridge, Mass.: Harvard University Press.
Crowder, Michael. 1967. *Sengal: A Study in French Assimilation Policy.* Rev. ed. London: Methuen.
Crozier, Brian. 1975. *The Soviet Presence in Somalia.* London: Institute for the Study of Conflict.
Daly, Vere T. 1975. *A Short History of the Guyanese People.* London: Macmillan.
Decalo, Samuel. 1976a. *Coups and Army Rule in Africa.* New Haven, Conn.: Yale University Press.
———. 1976b. *Historical Dictionary of Togo.* Metuchen, N.J.: Scarecrow Press.
———. 1977. *Historical Dictionary of Chad.* Metuchen, N.J.: Scarecrow Press.
Drysdale, John G. S. 1964. *The Somali Dispute.* London: Pall Mall Press.
Dupuy, Trevor N., Grace P. Hayes, and John A. C. Andrews. 1980. *The Almanac of World Military Power.* 4th ed. Novato, Calif.: Presidio Press.
Esman, Milton J. 1972. *Administration and Development in Malaysia.* Ithaca, N.Y.: Cornell University Press.
Ewing, Laurence L., and Robert C. Sellers, eds. 1966. *The Reference Handbook of the Armed Forces of the World.* Washington, D.C.: Sellers & Associates.
Farer, Tom J. 1976. *War Clouds on the Horn of Africa.* New York: Carnegie Endowment for International Peace.
Fidel, Kenneth, ed. 1975. *Militarism in Developing Countries.* New Brunswick, N.J.: Transaction Books.
Finer, Samuel E. 1962. *The Man on Horseback.* London: Pall Mall Press.
Fletcher, Nancy McHenry. 1969. *The Separation of Singapore from Malaysia.* Ithaca, N.Y.: Cornell University Press.
Foley, Charles, and W. I. Scobie. 1975. *The Struggle for Cyprus.* Stanford, Calif.: Hoover Institution Press.
Foran, W. R. 1962. *The Kenya Police, 1887-1960.* London: R. Hale.
Foray, Cyril P. 1977. *Historical Dictionary of Sierra Leone.* Metuchen, N.J.: Scarecrow Press.
Foster, Phillip and Aristide R. Zolberg, eds. 1971. *Ghana and the Ivory Coast: Perspectives on Modernization.* Chicago: University of Chicago Press.
Fraser, Lionel M. 1971. *History of Trinidad.* London: Cass.
Frederiksen, P. C., and Robert E. Looney. 1983. "Defense Expenditures and Economic Growth in Developing Countries," *Armed Forces and Society* 9.4 (Summer): 633.
Gauze, Rene. 1973. *The Politics of Congo Brazzaville,* tr. Virginia Thompson and Richard Adloff. Stanford, Calif.: Hoover Institution Press.
Gavin, R. J. 1975. *Aden Under British Rule, 1839-1967.* New York: Barnes & Noble.

Gerard-Libois, Jules. 1966. *Katanga Secession,* tr. Rebecca Young. Madison: University of Wisconsin Press.

Gerteiny, Alfred G. 1967. *Mauritania.* New York: Praeger.

Gochman, Charles S. 1975. "Status, Conflict, and War: The Major Powers, 1820-1970." Ph.D. dissertation, University of Michigan.

Goulet, Denis. 1978. *Looking at Guinea Bissau: A New Nation's Development Strategy.* Washington, D.C.: Overseas Development Council.

Gray, Colin S. 1971. "The Arms Race Phenomenon," *World Politics* 14.1 (Oct.): 39-79.

Great Britain, Colonial Office. 1954. "Report of the West African Forces Conference, Lagos, April 20-24, 1953." Colonial Report No. 304.

Grundy, Kenneth W. 1968. *Conflicting Images of the Military in Africa.* Nairobi: East African Publishing House.

Gukiina, Peter M. 1972. *Uganda: Case Study in Political Development.* Notre Dame, Ind.: University of Notre Dame Press.

Gutteridge, William. 1962. *Armed Forces in New States.* London: Oxford University Press.

———. 1965. *Military Institutions and Power in the New States.* London: Pall Mall Press.

———. 1969. *The Military in African Politics.* London: Methuen.

———. 1975. *Military Regimes in Africa.* London: Methuen.

Guyer, D. 1970. *Ghana and the Ivory Coast: The Impact of Colonialism in an African Setting.* New York: Exposition Press.

Harvey, David, ed. 1980. *Defense and Foreign Affairs Handbook.* Washington, D.C.: Copely & Associates.

Harwich, Christopher. 1961. *Red Dust: Memories of the Ugandan Police, 1935-55.* London: V. Stuart.

Headrick, Rita. 1978. "African Soldiers in World War II," *Armed Forces and Society* 4.3 (Spring): 501-26.

Holtham, Gerald, and Arthur Hazelwood. 1976. *Aid and Inequality: British Development Assistance to Kenya.* London: Croom Helm.

Howard, Michael. 1976. *War in European History.* London: Oxford University Press.

Hudson, W. J., ed. 1974. *New Guinea Empire: Australia's Colonial Experience.* Melbourne: Cassell.

Huntington, Samuel P. 1957. *The Soldier and the State.* Cambridge, Mass.: Harvard University Press.

———. 1971. "The Change to Change: Organization, Development and Politics," *Comparative Politics* 3.3 (April): 283-322.

International Institute for Strategic Studies. 1960- . *The Military Balance.* London: IISS. Annual.

Irving, Brian, ed. 1972. *Guyana: A Composite Monograph.* Hato Rey, Puerto Rico: Inter American University Press.

Jackman, Robert W. 1976. "Politicians in Uniform," *American Political Science Review* 1970 (Dec.): 1078-97.

Jackson, Henry F. 1977. *The FLN in Algeria: Party Development in a Revolutionary Society.* Westport, Conn.: Greenwood Press.

James, Harold, and Denis Sheilsmall. 1971. *The Undeclared War: The Story of the Indonesian Confrontation, 1962-1966.* London: Lep Cooper.

Janowitz, Morris. 1964. *The Military in the Political Development of New Nations*. Chicago: University of Chicago Press.
———. 1977. *Military Institutions and Coercion in the Developing Nations*. Chicago: University of Chicago Press.
Janowitz, Morris, and Jacques Van Doorn, eds. 1971. *On Military Intervention*. Rotterdam: Rotterdam University Press.
Johnson, J. 1972. *Econometric Methods*. 2d ed. New York: McGraw Hill.
Johnson, John J., ed. 1962. *The Role of the Military in Underdeveloped Countries*. Princeton, N.J.: Princeton University Press.
Johnson, Willard R. 1970. *The Cameroon Federation: Political Integration in a Fragmentary Society*. Princeton, N.J.: Princeton University Press.
Kalck, Pierre. 1971. *The Central African Republic: A Failure in Decolonization*, tr. Barbara Thompson. New York: Praeger.
Kanja, Thomas R. 1972. *Conflict in the Congo: The Rise and Fall of Lumumba*. Middlesex, Eng.: Penguin Books.
Kaplan, Irving, ed. 1979. *Angola, a Country Study*. 2d ed. Washington, D.C.: American University.
Kaplan, Irving, et al. 1977. *Area Handbook for Mozambique*. 2d ed. Washington, D.C.: U.S. Government Printing Office.
Kasper, Wolfgang. 1974. *Malaysia: A Study in Successful Economic Development*. Washington, D.C.: American Enterprise Institute.
Keegan, John, ed. 1979. *World Armies*. London: Macmillan.
Kelleher, Catherine M., ed. 1974. *Political Military Systems: Comparative Perspectives*. Beverly Hills, Calif.: Sage.
Kemp, Geoffrey. 1970. *Classification of Weapons Systems and Force Designs in Less Developed Country Environments*. Cambridge, Mass.: MIT Press.
Khaketla, B. Makalo. 1972. *Lesotho 1970: An African Coup Under the Microscope*. Berkeley: University of California Press.
Kimambo, Isaria N., and A. J. Teniu, eds. 1969. *A History of Tanzania*. Nairobi: East African Publishing House.
Kmenta, Jan. 1971. *Elements of Econometrics*. New York: Macmillan.
Krivinyi, Nikolas, ed. 1977. *World Military Aviation*. New York: Arco Publishing Co.
Kurtz, Laura S. 1978. *Historical Dictionary of Tanzania*. Metuchen, N.J.: Scarecrow Press.
Labayle Couhat, Jean. 1976- . *Combat Fleets of the World*. Annapolis, Md.: U.S. Naval Institute Press. Biennial.
Lee, J. M. 1969. *African Armies and Civil Order*. New York: Praeger.
Lefever, Ernest W. *Spear and Scepter*. 1970. Washington, D.C.: Brookings Institution.
Leith, J. Clark. 1974. *Ghana*. New York: Columbia University Press.
Lemarchand, René. 1964. *Political Awakening in the Belgian Congo*. Berkeley: University of California Press.
———. 1970. *Rwanda and Burundi*. London: Pall Mall Press.
———. 1974. *Selective Genocide in Burundi*. London: Methuen.
Levine, Victor T. 1964. *The Cameroons from Mandate to Independence*. Berkeley: University of California Press.
Lingoes, James C. 1973. *The Guttman-Lingoes Nonmetric Program Series*. Ann Arbor, Mich.: Mathesis Press.

Lobban, Richard. 1979. *Historical Dictionary of the Republics of Guinea Bissau and Cape Verde*. Metuchen, N.J.: Scarecrow Press.
Luckham, Robin. 1971. *The Nigerian Military*. Cambridge: Cambridge University Press.
McDonald, G. C. 1969. *Area Handbook for Burundi*. Washington, D.C.: U.S. Government Printing Office.
McFarland, Daniel Miles. 1978. *Historical Dictionary of Upper Volta*. Metuchen, N.J.: Scarecrow Press.
McNeill, William H. 1963. *The Rise of the West: A History of the Human Community*. Chicago: University of Chicago Press.
———. 1982. *The Pursuit of Power: Technology, Armed Force, and Society Since A.D. 1000*. Chicago: University of Chicago Press.
Mayer, Emerico S. 1968. *Ghana: Past and Present*. 2d ed. The Hague: Levisson Press.
Mazrui, Ali A. 1975. *Soldiers and Kinsmen in Uganda*. Beverly Hills, Calif.: Sage.
Melady, T. P. 1974. *Burundi: The Tragic Years*. Maryknoll, N.Y.: Orbis Press.
Merkle, Peter H. 1977. "The Study of European Political Development," *World Politics* 29.3 (April): 462-75.
Miller, N. J. 1971. *The Nigerian Army, 1956-1966*. London: Methuen.
Moore, J. E., ed. n.d. *Jane's Fighting Ships*. London: Jane's Yearbooks. Annual.
Morton, Kathryn. 1975. *Aid and Dependence: British Aid to Malawi*. London: Croom Helm.
Mowoe, Isaac James, ed. 1980. *The Performance of Soldiers as Governors: African Politics and the African Military*. Washington, D.C.: University Press of America.
Moyse-Bartlett, Hubert. 1956. *The King's African Rifles*. Aldershot, Eng.: Gale & Polden.
Munslow, Barry. 1983. *Mozambique: The Revolution and Its Origins*. London: Longman.
Nettleford, R. 1970. *Mirror, Mirror: Identity, Race and Protest in Jamaica*. Kingston: W. Collins & Sangster.
Nordlinger, Eric. 1970. "Soldiers in Mufti: The Impact of Military Rule Upon Economic and Social Change in the Non-Western States," *American Political Science Review* 64 (Dec.): 1131-48.
O'Ballance, Edgar. 1967. *The Algerian Insurrection, 1954-1962*. London: Faber.
Olorsunsola, Victor A. 1977. *Soldiers and Power: The Development Performance of the Nigerian Military Regime*. Stanford, Calif.: Hoover Institution Press.
Organski, A. F. K., and Jacek Kugler. 1980. *The War Ledger*. Chicago: University of Chicago Press.
O'Toole, Thomas E. 1978. *Historical Dictionary of Guinea*. Metuchen, N.J.: Scarecrow Press.
Ottaway, David, and Marina Ottaway. 1970. *Algeria: The Politics of a Socialist Revolution*. Berkeley: University of California Press.
Owen, John I. H., ed. 1976a. *NATO Infantry and Its Weapons*. Boulder, Col.: Westview.
———. 1976b. *Warsaw Pact Infantry and Its Weapons*. London: Brassey's.

Oxaal, Ivar. 1971. *Race and Revolutionary Consciousness in Trinidad and Tobago*. Cambridge, Mass.: Schenkman.

Pachai, Bridglal. 1973. *Malawi: The History of the Nation*. London: Longman.

Panikkar, Kavalam Madhusudan. *Revolution in Africa*. Westport, Conn.: Greenwood.

Pascal, James Imperato. 1975. *Historical Dictionary of Mali*. Metuchen, N.J.: Scarecrow Press.

Pettman, Jan. 1974. *Zambia: Security and Conflict*. Sussex, Eng.: Julian Friedmann.

Pike, John G. 1968. *Malawi: A Political and Economic History*. New York: Praeger.

Poltholm, Christian P. 1970. *Four African Political Systems*. Englewood Cliffs, N.J.: Prentice-Hall.

Polyviou, Polyvios G. 1980. *Cyprus, Conflict and Negotiation, 1960-1980*. London: Duckworth.

Pye, Lucien. 1965. "The Concept of Political Development," *Annals of the American Academy of Political and Social Science* 2.13 (March): 358-66.

Quandt, William B. 1969. *Revolution and Political Leadership: Algeria, 1954-1968*. Cambridge, Mass.: MIT Press.

Ra'anan, Uri, Robert L. Pfaltzgraff, Jr., and Geoffrey Kemp, eds. 1978. *Arms Transfers to the Third World: The Military Buildup in Less Industrial Countries*. Boulder, Col.: Westview.

Riviere, Claude. 1977. *Guinea: The Mobilization of the People*, tr. Virginia Thompson and Robert Adloff. Ithaca, N.Y.: Cornell University Press.

Robinson, Arthur N. R. 1971. *The Mechanics of Independence: Patterns of Political and Economic Transformation in Trinidad and Tobago*. Cambridge, Mass.: MIT Press.

Rokkan, Stein, and S. N. Eisenstadt, eds. 1973. *Building States and Nations*. Beverly Hills, Calif.: Sage.

Ronen, D. 1975. *Dahomey Between Tradition and Modernity*. Ithaca, N.Y.: Cornell University Press.

Ross, William. 1968. *Kenya from Within: A Short Political History*. London: Cass.

Rotberg, Robert I. 1975. *The Rise of Nationalism in Central Africa: The Making of Zambia and Malawi*. Cambridge, Mass.: Harvard University Press.

Russett, Bruce, ed. 1972. *War Peace and Numbers*. Beverly Hills, Calif.: Sage.

Schumacher, Edward J. 1975. *Politics, Bureaucracy, and Rural Development in Senegal*. Berkeley: University of California Press.

Serapiao, Luis B. 1979. *Mozambique in the Twentieth Century: From Colonialism to Independence*. Washington, D.C.: University Press of America.

Shepard, Roger N., A. Kimball Romney, and Sara Beth Nerlove, eds. 1972. *Multidimensional Scaling: Theory and Applications in the Behavioral Sciences*, Vol. 1. New York: Seminar Press.

Simon, Sheldon W., ed. 1978. *The Military and Security in the Third World: Domestic and International Impacts*. Boulder, Col.: Westview.

Sivard, Ruth Leger. 1974- . *World Military and Social Expenditures*. Leesburg, Va.: World Priorities.

Skurnik, W. A. E. 1972. *The Foreign Policy of Senegal.* Evanston, Ill.: Northwestern University Press.
Snyder, Frank Gregory. 1965. *One Party Government in Mali.* New Haven, Conn.: Yale University Press.
Spence, John Edward. 1968. *Lesotho: The Politics of Dependence.* London: Oxford University Press.
Stephens, Robert. 1966. *Cyprus, a Place of Arms.* London: Pall Mall Press.
Steven, Richard P. 1967. *Lesotho, Botswana, and Swaziland: The Former High Commissions Territories in Southern Africa.* New York: Praeger.
Stockholm International Peace Research Institute. 1969- . *SIPRI Yearbook of World Armaments and Disarmament.* Stockholm: Almqvist & Wiksell.
———. 1975. *Arms Trade Registers: The Arms Trade with the Third World.* Cambridge, Mass.: MIT Press.
Stoll, Richard J. 1977. "Conjoint Measurement." Mimeo. Ann Arbor, Mich.
Talbott, John F. 1980. *The War Without a Name: France in Algeria, 1954-1962.* New York: Knopf.
Taylor, J. W., ed. n.d. *Jane's All the World's Aircraft.* London: Jane's Yearbooks. Annual.
Thompson, Virginia M. 1972. *West Africa's Council of the Entente.* Ithaca, N.Y.: Cornell University Press.
Thompson, Virginia M., and Richard Adloff. 1972. *The Malagasy Republic.* Stanford, Calif.: Stanford University Press.
———. 1974. *Historical Dictionary of the Peoples' Republic of the Congo.* Metuchen, N.J.: Scarecrow Press.
Tilly, Charles, ed. 1975. *The Formation of National States in Western Europe.* Princeton, N.J.: Princeton University Press.
Tordoff, William. 1967. *Government and Politics in Tanzania.* Nairobi: East African Publishing House.
———, ed. 1974. *Politics in Zambia.* Manchester, Eng.: Manchester University Press.
U.S. Arms Control and Disarmament Agency. 1971-82. *World Military Expenditures and Arms Transfers.* Washington, D.C.: U.S. Government Printing Office. Annual.
Van Doorn, Jacques, ed. 1968. *Armed Forces and Society.* The Hague: Mouton.
Wallerstein, I. M. 1964. *The Road to Independence: Ghana and the Ivory Coast.* The Hague: Mouton.
Weinstein, Warren, ed. 1975. *Chinese and Soviet Aid to Africa.* New York: Praeger.
———. 1976a. *Historical Dictionary of Burundi.* Metuchen, N.J.: Scarecrow Press.
———. 1976b. *Political Conflict and Ethnic Strategies.* Syracuse, N.Y.: Syracuse University Press.
Welch, Claude E., Jr. 1970. *Soldier and State in Africa.* Evanston, Ill.: Northwestern University Press.
———. 1978. "Civil Military Relations in Newer Commonwealth States: The Transfer and Transformation of British Models," *Journal of Developing Areas* 12.2 (Jan.): 153-70.
———. 1983. "Military Disengagement from Politics: Lessons from West Africa," *Armed Forces and Society* 9.4 (Summer): 539.

———. 1985. "Civil-Military Relations: Perspectives from the Third World," *Armed Forces and Society* 11.2 (Winter): 183.

Westebbe, Richard M. 1971. *The Economy of Mauritania*. New York: Praeger.

———. 1974. *Chad: Development Potential and Constraints*. Washington, D.C.: World Bank.

Whynes, David K. 1979. *The Economics of Third World Military Expenditures*. Austin: University of Texas Press.

Wilairat, Kawin. 1975. *Singapore's Foreign Policy: The First Decade*. Singapore: Institute of Southeast Asian Studies.

Winstone, Harry V. F., and Zahra Freeth. 1972. *Kuwait: Prospect and Reality*. London: Allen & Unwin.

Winter, J. M., ed. 1975. *War and Economic Development*. Cambridge: Cambridge University Press.

World Bank. 1979. *Bangladesh: Current Trends and Development Issues*. Washington, D.C.

Wright, Carol. 1974. *Mauritius*. Harrisburg, Pa.: Stackpole Books.

Young, Crawford. 1965. *Politics in the Congo: Decolonization and Independence*. Princeton, N.J.: Princeton University Press.

Yu, George T. 1975. *China's African Policy: A Study of Tanzania*. New York: Praeger.

Zolberg, Aristide R. 1964. *One Party Government in the Ivory Coast*. Princeton, N.J.: Princeton University Press.

Index

Library of Congress Cataloging-in-Publication Data

Mullins, A. F., 1947-
Born arming.

(ISIS studies in international security and arms control)
Bibliography: p.
Includes index.
1. Developing countries—Defenses. 2. Munitions—
Developing countries. I. Title. II. Series.
UA10.M85 1987 355′.03301724 86-23043
ISBN 0-8047-1375-8 (alk. paper)